INTRODUCTION TO THE
PHYSICAL CHEMISTRY
OF FOODS

Christos Ritzoulis

Translated by Jonathan Rhoades

INTRODUCTION TO THE
PHYSICAL CHEMISTRY
OF FOODS

CRC Press
Taylor & Francis Group
Boca Raton   London   New York

CRC Press is an imprint of the
Taylor & Francis Group, an **informa** business

CRC Press
Taylor & Francis Group
6000 Broken Sound Parkway NW, Suite 300
Boca Raton, FL 33487-2742

© 2013 by Taylor & Francis Group, LLC
CRC Press is an imprint of Taylor & Francis Group, an Informa business

No claim to original U.S. Government works

Printed on acid-free paper
Version Date: 20130308

International Standard Book Number-13: 978-1-4665-1175-0 (Hardback)

Library of Congress Cataloging-in-Publication Data

Ritzoulis, Christos.
   Introduction to the physical chemistry of foods / author, Christos Ritzoulis.
   pages cm
   Summary: "This book provides an introduction to the physico-chemical basis and mathematical description of notions such as distillation, crystallization, sublimation, protein denaturation, emulsions, kinetics, enzymatic activity, and rheology. It lays the foundations in explaining the texture of ice cream, the foaming of beer, and the stability of milk, correlating fundamental physico-chemical concepts such as freezing, distillation, (re)crystallization, foaming, emulsification, aggregation and clarification to their applications in the fields of processing, microbiology and bioavailability of foods. Special attention has been paid so that the content is easily accessible by readers with basic background knowledge of physics, mathematics and chemistry"-- Provided by publisher.
   Includes bibliographical references and index.
   ISBN 978-1-4665-1175-0 (hardback)
   1. Food--Analysis. 2. Food--Composition. 3. Chemistry, Technical. I. Title.

   TX541.R58 2013
   664'.07--dc23   2012050936

Visit the Taylor & Francis Web site at
http://www.taylorandfrancis.com

and the CRC Press Web site at
http://www.crcpress.com

Contents

Introduction to the Greek edition..ix
Preface to the English edition..xi
About the author xiii

Chapter 1   The physical basis of chemistry...1
1.1   Thermodynamic systems 1
1.2   Temperature 2
1.3   Deviations from ideal behavior: Compressibility............................4
      1.3.1   van der Waals equation 6
      1.3.2   Virial equation..9

Chapter 2   Chemical thermodynamics...13
2.1   A step beyond temperature..13
2.2   Thermochemistry 16
2.3   Entropy..17
2.4   Phase transitions...21
2.5   Crystallization...27
2.6   Application of phase transitions: Melting, solidifying, and
      crystallization of fats...27
      2.6.1   Chocolate: The example of cocoa butter..............................30
2.7   Chemical potential 31

Chapter 3   The thermodynamics of solutions.............................35
3.1   From ideal gases to ideal solutions 35
3.2   Fractional distillation 38
3.3   Chemical equilibrium 41
3.4   Chemical equilibrium in solutions 44
3.5   Ideal solutions: The chemical potential approach 46
3.6   Depression of the freezing point and elevation of the boiling
      point..47
3.7   Osmotic pressure 48
3.8   Polarity and dipole moment..50
      3.8.1   Polarity and structure: Application to proteins 51

3.9   Real solutions: Activity and ionic strength...................................52
3.10 On pH: Acids, bases, and buffer solutions...............................53
3.11 Macromolecules in solution 57
3.12 Enter a polymer...58
3.13 Is it necessary to study macromolecules in food and
      biological systems in general?...59
      3.13.1 Intrinsic viscosity..60
3.14 Flory–Huggins theory of polymer solutions 60
      3.14.1 Conformational entropy and entropy of mixing.............. 61
      3.14.2 Enthalpy of mixing...66
      3.14.3 Gibbs free energy of mixing 67
3.15 Osmotic pressure of solutions of macromolecules 68
      3.15.1 The Donnan effect 68
3.16 Concentrated polymer solutions 69
3.17 Phase separation 70
      3.17.1 Phase separation in two-solute systems............................72

Chapter 4   Surface activity...77
4.1   Surface tension 77
4.2   Interface tension..79
      4.2.1 A special extended case...80
4.3   Geometry of the liquid surface: Capillary effects.............................81
4.4   Definition of the interface...82
4.5   Surface activity...83
4.6   Adsorption...85
      4.6.1 Thermodynamic basis of adsorption.................................85
      4.6.2 Adsorption isotherms 85
4.7   Surfactants 90

Chapter 5   Surface-active materials 93
5.1   What are they, and where are they found? 93
5.2   Micelles 94
5.3   Hydrophilic-lipophilic balance (HLB), critical micelle
      concentration (cmc), and Krafft point...96
5.4   Deviations from the spherical micelle..98
5.5   The thermodynamics of self-assembly.. 100
5.6   Structures resulting from self-assembly 104
      5.6.1 Spherical micelles 107
      5.6.2 Cylindrical micelles.. 107
      5.6.3 Lamellae: Membranes... 108
      5.6.4 Hollow micelles.. 109
      5.6.5 Inverse structures 110
5.7   Phase diagrams... 112

5.8   Self-assembly of macromolecules: The example of proteins......... 112
        5.8.1   Why are all proteins not compact spheres with their
                 few nonpolar amino acids on the inside?....................114
        5.8.2   How do proteins behave in solution? 114
        5.8.3   A protein folding on its own: The Levinthal paradox 116
        5.8.4   What happens when proteins are heated?...................117
        5.8.5   What is the effect of a solvent on a protein? 118
        5.8.6   What are the effects of a protein on its solvent?........119
        5.8.7   Protein denaturation: An overview 120
        5.8.8   Casein: Structure, self-assembly, and adsorption............ 121
        5.8.9   Adsorption and self-assembly at an interface: A
                 complex example 122
        5.8.10   To what extent does the above model apply
                 to the adsorption of a typical spherical protein? 123
        5.8.11   Under what conditions does a protein adsorb to a
                 surface, and how easily does it stay adsorbed there?....... 124

Chapter 6   Emulsions and foams..127
6.1   Colloidal systems... 127
        6.1.1   Emulsions and foams nomenclature 128
6.2   Thermodynamic considerations.................................. 130
6.3   A brief guide to atom-scale interactions 131
        6.3.1   van der Waals forces 131
        6.3.2   Hydrogen bonds 133
        6.3.3   Electrostatic interactions............................. 134
        6.3.4   DLVO theory: Electrostatic stabilization of colloids......... 135
        6.3.5   Solvation interactions................................ 137
        6.3.6   Stereochemical interactions: Excluded volume forces 138
6.4   Emulsification...141
        6.4.1   Detergents: The archetypal emulsifiers.................. 144
6.5   Foaming 145
6.6   Light scattering from colloids................................... 146
6.7   Destabilization of emulsions and foams.......................... 147
        6.7.1   Gravitational separation: Creaming 148
        6.7.2   Aggregation and flocculation 150
        6.7.3   Coalescence.. 152
        6.7.4   Phase inversion 153
        6.7.5   Disproportionation and Ostwald ripening................ 153

Chapter 7   Rheology...157
7.1   Does everything flow? 157
7.2   Elastic behavior: Hooke's law 159
7.3   Viscous behavior: Newtonian flow 161

7.4   Non-Newtonian flow 162
        7.4.1   Time-independent non-Newtonian flow 162
        7.4.2   Time-dependent non-Newtonian flow 164
7.5   Complex rheological behaviors 165
        7.5.1   Application of non-Newtonian flow: Rheology of
                 emulsions and foams 165
7.6   How does a gel flow? (Viscoelasticity) 168
7.7   Methods for determining viscoelasticity 168
        7.7.1   Creep 168
        7.7.2   Relaxation 169
        7.7.3   Dynamic measurements: Oscillation............................. 169

Chapter 8   Elements of chemical kinetics............................. 173
8.1   Diamonds are forever?.. 173
8.2   Concerning velocity 174
8.3   Reaction laws... 174
8.4   Zero-order reactions.. 176
8.5   First-order reactions 177
        8.5.1   Inversion of sucrose... 178
8.6   Second- and higher-order reactions.................................. 180
8.7   Dependence of velocity on temperature 182
8.8   Catalysis 183
8.9   Biocatalysts: Enzymes... 184
8.10   The kinetics of enzymic reactions................................... 185
        8.10.1   Lineweaver–Burk and Eadie–Hofstee graphs 187

Bibliography.. 191
Index 195

Introduction to the Greek edition

The driving force for writing the present book is the current absence of a text that, starting from the principles of physical chemistry (a demanding science), will end up in the description of food behavior in physicochemical terms. The final text should be concise and easy to absorb, but without being over-simplified.

Written on the basis of my teaching and research experience in the field of physical chemistry of foods, I hope that this text provides the necessary depth and mathematical completeness, without sacrificing simplicity and directness of presentation. When written, this book was aimed at undergraduate and postgraduate students and young researchers working in the field of food. However, I believe that it can be equally useful to students, researchers, and professionals in nearby fields such as the pharmaceutical and health sciences, and cosmetics and detergent technology.

At many points in the text, new terms had to be introduced, for which, to the best information of this author, no appropriate words exist in Greek. Thus, for example, the term κροκίδωση από εκκένωση renders what is known in English as "depletion flocculation," while the terms ωρίμανση κατά Ostwald and δυσαναλογία are used for "Ostwald ripening" and "disproportionation," respectively. It is self-explanatory that proposals for the amelioration of the novel terms are welcome.

Despite the painstaking and repeated checks of the text, unavoidably some spelling, syntax, or arithmetical errors might have escaped attention. It is the strong wish of the author that the readers point out such errata, as well as any unclear parts in the text.

I would like to thank Professor Stylianos Raphaelides, Professor George Ritzoulis, and Dr. Chrisi Vasiliadou for the time they devoted to reading the chapters and their useful propositions for corrections in the text.

Christos Ritzoulis
Thessaloniki

Preface to the English edition

This book is aimed at introducing the basic concepts of physical chemistry to postgraduate and undergarduate students and to scientists and engineers who have an interest in the field of foods, but also in the neighboring fields of pharmaceuticals, materials, and cosmetics. The rationale behind this book is to start from basic physics and chemistry, and then build up the reader's understanding of those parts of physical chemistry (a separate science in its own right) directly related to food, including processes of crystallization, melting, distillation, blanching, homogenization, and properties as diverse as rheology, color, and foam stability.

Chapter 1 introduces the basic physicochemical entity, which is the ideal gas, along with the concept of temperature, followed by a description of the real gases in terms of deviations from ideal behavior. Chapter 2 carries on with a discussion of the Second Law of Thermodynamics, and describes the formation of liquids and solids along with the relevant phase transitions. Chapter 3 continues the discussion, dealing with the properties of solutions of small molecules and of polymers. Then, Chapter 4 introduces the notion of surface activity, defines the surface/interface and the adsorption of molecules, and introduces surface-active molecules. Chapter 5 discusses the properties of amphiphilic molecules, with an emphasis on self-assembled and colloidal structures, followed by relevant examples from the field of food proteins. Chapter 6 discusses colloidal entities focused on emulsions and foams, and Chapter 7 introduces the main macroscopic manifestations of colloidal (and other) interactions in terms of rheology. Then finally, Chapter 8 deals with the science of chemical/enzymic kinetics, a recurrent theme in the study of foods.

Here, I must thank Dr. Jonathan Rhoades for the excellent work in translating the original Greek text. Apart from his philological task, Jon brought forward many comments and remarks of a scientific nature that clarified and ameliorated the final result. I would also like to thank the people at CRC Press for their expert professional and kind assistance throughout this project.

Christos Ritzoulis
Thessaloniki

About the author

Christos Ritzoulis studied chemistry at the Aristotle University of Thessaloniki, and food science (M.Sc. and Ph.D.) at the University of Leeds. He has worked as a postdoctoral researcher at the Department of Chemical Engineering of the Aristotle University of Thessaloniki, and as an analyst at the Hellenic States General Chemical Laboratories. Today, Christos is Senior Lecturer of Food Chemistry at the Department of Food Technology at TEI of Thessaloniki, where he teaches food chemistry and physical chemistry of foods.

chapter one

The physical basis of chemistry

1.1   Thermodynamic systems

In physical chemistry, the term *system* refers to a clearly defined section of the universe that is separated from the remainder of the universe by a *boundary*. The region outside the system that is in immediate contact with the boundary is called the *environment* or the *surroundings* of the system. A system is described as *open* if both material and energy can pass between the system and the environment, *closed* if only energy but not material can pass across the boundary, and *isolated* if neither energy nor material can enter or leave the system. Boundaries can be described as *permeable* if material can pass in both directions, *semipermeable* if material can pass in one direction only, or *adiabatic* if neither matter nor energy can traverse the boundary.

Another way of distinguishing between systems is their classification into *homogeneous* and *heterogeneous systems*. A homogeneous system has the same composition and properties throughout. Such systems are said to comprise only one phase, and are thus termed *monophasic*. In contrast, heterogeneous systems are composed of more than one phase. The concept of homogeneity is related to the scale on which we consider the material of the system, and this must always be defined. For example, milk that is homogeneous on a macroscopic scale consists of a heterogeneous colloidal suspension of fat droplets and proteins when examined at the micrometer level. Similarly, a clear "homogeneous" gel consists, at the scale of tens or hundreds of nanometers, of two distinct phases: water and hydrated polysaccharides.

A thermodynamic system is described using three basic parameters: the pressure (P), the volume (V), and the temperature of the system (T). A system that does not change over time (a system in *equilibrium*) has a fixed value for each of these parameters, which together define the *thermodynamic state* of the system. These three parameters are sufficient to describe at least the simplest material, a gas of which the molecules move freely in any direction. The molecules, of total number n, are considered to be without volume themselves and moving within a space of volume V, exerting pressure P on the walls of the container. Their average velocity is immediately related to the temperature T, with higher temperatures correlating to greater motility of the molecules. A gas in which the molecules can be considered to be of zero volume and noninteractive is described as an *ideal gas*, and forms the basis of the mathematical description of thermodynamic systems.

1.2   Temperature

For an ideal gas, the concept of temperature is inseparably bound to the pressure P and the volume V. For a given pair of values P and V, a number of molecules n has a given temperature T, notwithstanding the chemical composition of the gas. Let us say that the volume of an ideal gas is altered. This will lead to the alteration of the other two dependent variables P and T. The relationship between volume and pressure for a given temperature (or series of temperatures) and a constant value of n may be presented in a diagram such as that in Figure 1.1, which is known as a Clapeyron diagram.

The plot lines in Figure 1.1, referred to as *isotherms* as they represent points of the same temperature, enable the determination of the possible combinations of pressure and volume of an ideal gas at a given temperature. The gradient of the isotherms at temperature T_α is proportional to the ratio of pressure to volume at that point.

$$\frac{dP}{dV} = \frac{d\left(\frac{nRT_\alpha}{V}\right)}{dV} = -\frac{nRT_\alpha}{V^2} = -\frac{PV}{V^2} = -\frac{P}{V} \tag{1.1}$$

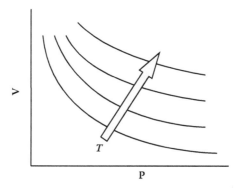

Figure 1.1 Typical pressure and volume curves for isothermic transformation. Every curve represents the solutions of Equation (1.1) for a specific temperature.

In the case of isothermic change, such as the slow thermosted compression of a low-pressure gas, at every point the plot line has a slope equal to the negative ratio of the pressure to the volume.

Now consider a transformation in which the volume V_α of a gas remains constant while its pressure is altered (known as an *isochoric* change). In this case, the equation of state for ideal gases can be written as

$$PV_\alpha = nRT \Rightarrow P = \left(\frac{n}{V_\alpha}R\right)T = (C_\alpha R)T \qquad (1.2)$$

Considering the contents of the parenthesis in the above equation ($C_\alpha R$) to be a constant value, it is apparent that the temperature is proportional to the pressure. The proportionality constant for an ideal gas is a product of the concentration of the gas C_α (mol dm^{-3}) and the constant R, which is the universal gas constant. On a plot of P against T, the value of R (~8.31441 J K^{-1} mol^{-1}) can be derived from the gradient of the plot line if C_α has unit value. If the line is extrapolated back to the point of zero pressure, the temperature value at the intercept is also zero Kelvin (0K). This value is known as absolute zero, and is between $-273.15°C$ and $-273.16°C$. Absolute zero is the starting point of the Kelvin scale, which is always used in thermodynamics rather than the Celsius scale. In a similar way, isobaric transformation can be defined as a change in volume with the pressure P_a remaining constant. In this case, a graph can be plotted of volume against temperature, with the gradient of the plot line (nR/P_a) dependent on the quantity of gas (mol) under pressure P_a.

Based on the above, is it possible to define a scale for an abstract concept such as temperature? Let us consider an ideal gas that undergoes isobaric (i.e., under constant pressure) heating to a temperature T. Its volume V_T at this temperature is given by

$$V_T = kT \qquad (1.3)$$

The volume is linear in relation to the temperature. If the same experiment is repeated for other pressures, straight plot lines would result that intercept the temperature axis at the same point—a lower temperature than that point would mean that the volume would become negative, which is clearly nonsensical. Thus, the point of intersection with the temperature axis gives the lowest temperature that is feasible—the aforementioned absolute zero (Figure 1.2).

In this way, the temperature scale can start to be defined as the solution for the pair of values P and V of the equation of state for ideal gases for a series of values of T.

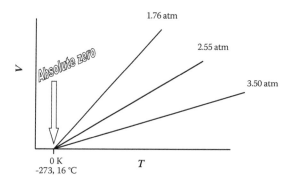

Figure 1.2 Typical temperature and volume plot lines for isobaric transformation. Every straight line plots the solutions of Equation (1.3) for a specific pressure.

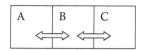

Figure 1.3 A series of three systems in which A is in contact and thermal equilibrium with B, and B is in contact and thermal equilibrium with C. The "zeroth law" states that A must therefore be in equilibrium with C.

Now consider three systems, A, B and C, of which A is in equilibrium with B, B is in equilibrium with A and C, and C is in equilibrium with B (Figure 1.3). Common experience dictates that by induction A and C must likewise be in equilibrium with each other. This empirical conclusion is considered axiomatic and accepted by thermodynamics as the "*zeroth*" *law*.

1.3   *Deviations from ideal behavior: Compressibility*

Throughout the preceding paragraphs, it was emphasized that the ideal gas equation assumes that the molecules are infinitely small and non-interactive. Therefore, in an ideal gas that occupies volume V_0, all of the volume is considered empty space. In addition, no form of force or mutual interaction between the molecules is predicted. Of course, in reality all atoms or molecules have a finite volume V_{mol} that can be ignored only if the concentration of atoms or molecules is very low. In that particular case, the volume of the gas V that is available to the molecules is $V = V_0 \gg \Sigma V_{mol} \sim 0$. However, when real molecules with volume occupy a space in the system in which they are distributed, then the space that they occupy is excluded from the other molecules. In these circumstances, for every molecule that is placed in a space of volume V_0 with a number n of similar molecules, the space that is actually available is $V = V_0 - b$, where b corresponds to the product of the volume V_{mol} that an individual molecule

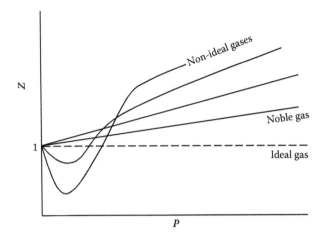

Figure 1.4 The relationship between the compressibility coefficient Z and pressure for a selection of gases. Note that the gases that do not exhibit strong mutual interactions (e.g., the noble gas He) are close to ideal gases in their behavior.

occupies and the number of molecules n in the system. For the purposes of volume calculations, molecules can be considered spheres. When two spheres approach one another, their centers cannot pass beyond the point that is determined by the sum of the radii of the two spheres (Figure 1.4).

The volume of the hypothetical sphere that constitutes the excluded volume for a pair of molecules is equal to $4/3\,\pi\,(2r)^3 = 32/3\,\pi r^3$, where r is the radius of the real sphere in question, in this case the molecule of gas. For a single molecule, the excluded volume is $1/2\,(32/3\,\pi r^3) = 16/3\,\pi r^3$.

The real volume V_{mol} of a molecule of radius r is equal to $4/3\,\pi r^3$, so, according to the above, the excluded volume b is four times the total volume that is occupied by all the molecules of a gas.

In addition, the ideal gas equation does not take account of any other mutual interactions—attractive or repulsive—that may occur between the gas molecules. An increase in attractive interactions limits the motility of the individual molecules of gas, causing a resistance to the flow of material, known as viscosity, and a transition to the liquid state. When the forces between molecules are sufficiently powerful, they can almost completely curtail the movement of the molecules, organizing them into structures that, in contrast to gases, are static. Such structures, with clear spatial arrangement of their constituent molecules or atoms, are solids.[*] Mutual interactions between the structural elements of a material will be extensively discussed in subsequent chapters.

---

[*] This is only one of the many definitions of solids, liquids, and gases. Other definitions will follow in later chapters where these concepts will be examined from other angles.

It may be said then that real gases approach the behavior of ideal gases when they occur in very low concentrations so that the average distance between molecules is very large and the probability that they will come into contact very small. In other cases, the compressibility factor Z = (PV)/RT is useful for the determination of deviation from ideal behavior. For a unit quantity (1 mole) of an ideal gas, Z = 1 (PV = RT). As the pressure of a real gas approaches zero, the value of Z approaches unity. At low pressures, a gas can have a value of Z below 1 (as is seen in the plot lines for CO_2 and O_2 in Figure 1.4); that is, it is more compressible than an ideal gas. This is due to attractive forces between the molecules that are not accounted for by the ideal gas laws.

1.3.1   van der Waals equation

At high pressures, the value of the compressibility factor is always greater than unity. This is due to the fact that at high pressures, the molecules come into contact with each other more frequently and the excluded volume b plays a more important role, reducing the total volume that is available to the molecules in the system (Figure 1.5). As a result, if we wish to calculate the parameters of a real gas, account must be taken of the volume occupied by the gas molecules $4\Sigma V_{mol} = b$ discussed previously (covolume or excluded volume) and the pressure that results from the mutual interactions between the molecules $p = a/V^2$ (a is a measure of the attraction between two molecules). Adjusting the terms for pressure and volume, the ideal gas equation can be rewritten (for $n = 1$ mol) as

$$\left(P + \frac{n^2 a}{V^2}\right)(V - nb) = nRT \tag{1.4}$$

This is referred to as the *van der Waals equation*. The covolume b and the intermolecular attraction coefficient a are unique for every gas. At this stage, perhaps the term "gas" should not be used, as sufficiently high values of the intermolecular attraction coefficient a give rise to liquid systems.

If the van der Waals equation is solved for 1 mol gas (let V_m equal the volume occupied by 1 mol), we have

$$V_m^3 - \left(b + \frac{RT}{P}\right)V_m^2 + \frac{\alpha}{P}V_m - \frac{ab}{P} = 0 \tag{1.5}$$

because a and b are direct functions of pressure and temperature, Equation (1.5) has three solutions for V_m for each set of P and T values. In

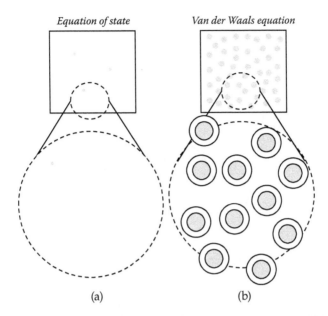

Figure 1.5 Schematic representation of the molecules of a gas as interpreted by (a) the ideal gas equation (left) and (b) the van der Waals equation (right).

practice, for increasing temperature, a *P-versus-V* diagram appears as in Figure 1.6.

The reader will note that above a temperature T_c, which is the *critical temperature* of the gas in question, only one solution to the van der Waals equation exists. We show later on that gases cannot be liquefied above this temperature. For temperatures below T_c, only the two extreme solutions of Equation (1.5) have physical significance: That which equates to the smallest volume represents the liquid phase, while that which equates to the largest volume represents the gaseous phase. As a gas is compressed, these isotherms show the continual change of volume during the compression. In real systems, the transition from gas to liquid phase is discontinuous with the compression of the gas, and its modeling cannot be satisfactorily approached with the van der Waals equation. In reality, the inflections shown in Figure 1.7 correspond to metastable states; that is, during compression from point A to point B (Figure 1.6), a supersaturated vapor will form, spontaneously phase separating into gas and liquid phases. In experimental measurements, usually a straight-line section runs parallel to the volume axis and connects the two extreme solutions of the equation.

As can be seen from Figure 1.7, when the volume of a gas under compression reduces to the first solution of the equation (point A), then

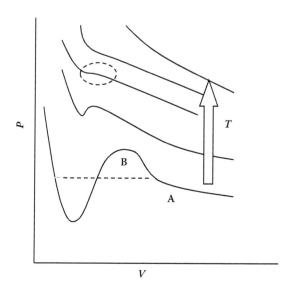

Figure 1.6 Solutions of the van der Waals equation. Every plot line presents the solutions for one temperature. Note that above a particular pressure, only one value of the pair $P - V$ is defined for each pressure (the formula becomes "one to one").

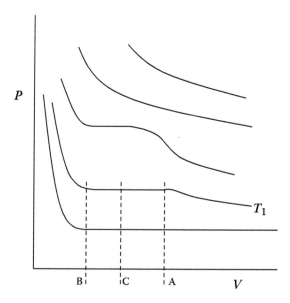

Figure 1.7 Isotherms of real gases. Note the shape that the triple solutions to the van der Waals equation have (compare with Figure 1.6).

condensation occurs immediately (throughout the length of the straight-line section AB) and without a reduction in pressure. The value of this pressure that remains stable until the volume reaches the other extreme value B (liquefaction) is called the vapor pressure. For every volume V, the quantity of gas that has been liquefied M_{liq} in relation to the mass that remains as gas M_{gas} is given by the relative sizes of the straight-line sections.

$$\frac{M_{liq}}{M_{gas}} = \frac{CA}{BC} \tag{1.6}$$

The critical data are associated with the van der Waals constants with the following formulae (in which the subscript marker c indicates critical quantity).

$$V_c = 3b \tag{1.7}$$

$$P_c = \frac{a}{27b^2} \tag{1.8}$$

$$T_c = \frac{8}{27}\frac{a}{Rb} \tag{1.9}$$

$$\frac{P_c V_c}{RT_c} = \frac{3}{8} \tag{1.10}$$

For real gases, the van der Waals equation may be written as

$$\left(P_{rel} + \frac{3}{V_{rel}^2}\right)(3V_{rel} - 1) = 8T_{rel} \tag{1.11}$$

where P_{rel}, V_{rel}, and T_{rel} are equivalent to the values P/P_c, V_m/V_c, and T/T_c.

1.3.2 Virial equation

It is clear that for a non-ideal gas ($Z \neq 1$), the Boyle–Mariotte equation of state is not valid, as the product PV is not constant. In this case, the compressibility factor for 1 mol of gas has the form proposed by the Kamerlingh–Onnes equation:

$$Z = \frac{PV}{RT} = 1 + \frac{B}{V} + \frac{C}{V^2} + \cdots \tag{1.12}$$

where B is the first virial coefficient that concerns mutual interactions between neighboring molecules. In this virial equation, the first virial coefficient (1) derives directly from the ideal gas equation of state in which molecules are assumed not to interact with each other; the second virial coefficient (B) relates to the interaction between two neighboring molecules; the third virial coefficient (C) relates to the mutual interaction between three neighboring molecules, and so on. For a particular gas, these coefficients depend only on the temperature and not on the pressure. The virial equation is valid until the gas is compressed into a liquid.

The attentive reader will note that because the virial coefficients B, C, etc., are inversely proportional to the powers of the volume, their significance is diminished from the third factor and onward.* The calculation of the second power coefficient is extremely useful for the determination of the forces between two molecules, especially in polymer science.

EXERCISES

1.1 How much pressure is exerted by 1 mol of an ideal gas in a closed container of volume 5 L at 20°C with a compressibility coefficient of 1.2? $R = 8.314$ J K^{-1}mol^{-1}.

Solution: Apply the ideal gas equation of state as modified by the compressibility coefficient.

1.2 4 mol of ideal gas are inserted into a system comprised of two identical spherical containers connected with a tube of negligible volume. When both tubes are at 27°C, the pressure is equal to 1 atm. Find the pressure and the number of mol in each one of the containers if the temperature of the one rises to 127°C, while the other's remains at 27°C. Consider that the pressure is the same in both containers.

Solution: Calculate the volume of the containers; after heating, as the volumes and pressures remain the same for both containers, write the equations for ideal gases for both containers, x mol for the one, $4 - x$ mol for the other; divide in parts; calculate the moles and then the pressures.

1.3 Calculate the number of molecules of O_2 that are contained in 3 L of atmospheric gas (79% v/v N_2, 21% v/v O_2) at the summit of Mount Everest (approximately 0.3 atm) and on a Mediterranean beach (approximately 1 atm) in June (–15°C on Everest, 35°C on the beach)

---

* From the point of view of statistical mechanics, this means that the probability of four molecules colliding simultaneously is extremely small.

and in January (–35°C on Everest, 5°C on the beach). Consider the gases to be ideal.

Solution: Apply the ideal gas equation of state. Pay attention to the units!

1.4 Pure H_2O_2 decomposes to water and oxygen. What will be the volume of O_2 released (assuming ideal behavior) from the decomposition of 2 L of H_2O_2 solution of density 1.447g cm^{-3}? Assuming that all of the produced oxygen is released as gas, calculate the pressure needed to bring the volume to 1 L. For the compression, assume a real gas with van der Waals constants α = 1.35 atm L^2 mol^{-2}; β = 0.0318 L mol^{-1}. Assume atomic masses to be 16 for O and 1 for H, atmospheric pressure and temperature equal to 20°C.

Solution: Calculate the mass of pure H_2O_2 in 2 L. Write equation $H_2O_2 \rightarrow H_2O + \frac{1}{2}O_2$. Calculate the number of mol O_2 produced. Using the gas equation of state, calculate their volume. Then assume $V = 1$ L and apply the van der Waals equation.

1.5 A 50-L cylinder contains 20 kg gas of molecular mass 14. If the technical specifications of the cylinder place a safety limit of 350 atm on the pressure, can it be stored in an area where the temperature varies from 50°C to 70°C? Note that the critical pressure of the gas is 35 atm bar and the critical temperature is –147°C.

Solution: Use Equations (1.10), (1.9), and (1.7) to calculate V_c, b, and a. Calculate the mol of gas and apply the van der Waals equation.

chapter two

Chemical thermodynamics

2.1 A step beyond temperature

Thermodynamics is the name given to the study of the energetic state of matter at rest. It is not concerned with the development of a phenomenon with time (kinetics), but concerns only the initial and final states of a transformation and only systems in equilibrium (i.e., systems with no tendency to further change). It constitutes a very powerful mathematical approach to physical and chemical problems as it allows quantitative calculations of transitions that take place without requiring consideration of the molecular composition of the system under study.

The amount of heat required to raise the temperature of a substance by 1 degree Kelvin is called the *heat capacity* of the substance. For gases, if the pressure remains constant during the heating process, the heat capacity is represented by the symbol C_P, while if the volume remains constant the symbol used is C_V. The corresponding symbols for molar heat capacity (i.e., the heat capacity of 1 mol of the substance) are c_p for constant pressure and c_V for constant volume.

Let us imagine that we raise the temperature of a gas by one degree Kelvin in two different ways: (1) by maintaining constant pressure and (2) by maintaining constant volume. If we had a means of measuring the quantity of heat that is imparted to the gas, we would observe that the heat capacity C_P is greater than C_V. The difference is due to the fact that in the first case, the gas changes its volume (expands). For this expansion, the gas is required to do work w, the energy for which must clearly come from the heat q that was supplied. The heat, then, can be transformed into work, which means that the *heat is a form of energy*. More specifically, the energy that relates to the *random* movement of the system components is called heat (q) and to it is attributed the thermal movement of atoms, molecules, crystals, and colloidal bodies. *Work* (w) is the energy related to the *organized* movement of the system components (relocation of center of gravity, flow of electrons, etc.). Organized movement such as this results in changes in system volume, and that leads to the mathematical expression for the work provided into a system:

$$w = -\int P dV \tag{2.1}$$

The sum of the energy and work gives the total energy of the system,* which is called the *internal energy U.*

A basic governing principle of thermodynamics is that the arithmetic sum of all the energetic transformations of an isolated system is equal to zero. This statement is called the First Law of Thermodynamics or the Principle of Conservation of Energy. This concept means that energy can be transformed from one form to another, but cannot be created from nothing or destroyed. In addition, every transfer of energy into or out of a system causes an equivalent change in internal energy U of the system.

For a change of state of a system from A to B, we can represent the change in internal energy U as

$$\Delta U = U_B - U_A = q + w \tag{2.2}$$

That is, the change in internal energy (final minus initial) is equal to the heat q that was absorbed by the system plus the work w done on the system. The change ΔU refers only to the initial state A and the final state B of the system, and not to any intermediate states through which it may have passed.[†]

For a small change in energy, we can write

$$\Delta U = \Delta q + \Delta w \tag{2.3}$$

or for small quantities of energy,

$$\delta U = \delta q + \delta w$$
$$\Delta U = \Delta q - \int P dV \tag{2.4}$$

Here, two scenarios can be distinguished:

1. If the volume of the system at the end of the transformation is the same as that at the start (e.g., it is enclosed within a container of fixed volume, $dV = 0$), then Equation (2.4) can be written as

* According to Einstein's well-known law $E = mc^2$, where c is the speed of light, the mass itself m can be converted to energy E. In basic thermodynamics, we work on the assumption that this does not happen, which is in accordance with our experience. In addition, the value of the internal energy U of a system is very difficult to determine. This is generally not a problem: the value U is not important, but ΔU is, that is, the *changes* in its value.
† If it were possible to go from A to B via two different pathways, each with a different value of ΔU, then a circular reaction pathway would be possible from A to B and back to A with $\Delta U \neq 0$; that is, energy would be produced from nothing.

$$\Delta U = q_v \tag{2.5}$$

where q_v is the heat under constant volume.

2. If the transformation is under constant pressure, $\int P dV = P(V_{final} - V_{initial}) = P\Delta V$, then Equation (2.4) can be written as

$$\Delta U = q_p - P\Delta V \tag{2.6}$$

where q_P is the change in heat under constant pressure. Let us call this heat q_P "enthalpy" and represent it with the symbol H for ease of reference. The enthalpy of a system under constant pressure can thus be defined by a transformation of Equation (2.6):

$$\Delta H = q_p = \Delta U + P\Delta V \tag{2.7}$$

and by extension

$$H = U + PV \tag{2.8}$$

The enthalpy here encompasses the internal energy U, namely the energy that is bound in the system, plus a parameter PV that is related to the pressure and the volume of the system.

For systems in which the initial and final pressures are the same, enthalpy is usually preferred as it includes within it any change in volume. Thus, in the description of a hypothetical explosion that involves the instantaneous generation of n new molecules of ideal gas that expand by volume ΔV (until the final pressure becomes equal to the initial), we have from the ideal gas equation that

$$P\Delta V = \Delta n \,(RT) \tag{2.9}$$

and from Equation (2.8)

$$\Delta H = \Delta U + \Delta n \,(RT) \tag{2.10}$$

On the contrary, for reactions where the change in volume of the products is negligible in comparison to the volume of the solvent in which the reaction occurs (a typical scenario in food and pharmaceutical sciences), we have $\Delta n = 0$ and Equation (2.10) becomes

$$\Delta H = \Delta U \tag{2.11}$$

2.2   Thermochemistry

Thermochemistry is the application of the First Law of Thermodynamics for the purpose of studying and quantifying the energy changes that take place during chemical reactions. Chemical reactions are divided into *endothermic* and *exothermic* reactions. If the temperature increases during a chemical reaction, heat will be transferred to the environment in order to restore the initial temperature of the system. This heat transfer is negative from the point of view of the system (i.e., heat is lost from the system to the environment), and such a reaction is termed *exothermic*. If the temperature decreases during the reaction, heat is transferred from the environment into the system. This is a positive heat transfer and the reaction is termed *endothermic*. In the case that heat cannot be exchanged with the environment (i.e., adiabatic conditions), endothermic reactions lower the temperature of the system and exothermic reactions increase it.

If the reaction takes place in a constant volume (e.g., within a sealed cylinder), the work done is equal to zero. In that case, from the First Law of Thermodynamics the following equation applies:

$$\Delta U = q_V \tag{2.12a}$$

Consider a reaction A $\rightarrow$ B. If the reaction takes place under constant pressure, as occurs in the majority of cases in the laboratory or in nature, then from the First Law of Thermodynamics, the following applies:

$$q_p = \Delta H \tag{2.12b}$$

In this equation, H symbolizes the *enthalpy*. Different types of enthalpy can be defined, depending on the reaction. The heat absorbed during a reaction under isobaric conditions (constant pressure) is defined as the *enthalpy of reaction*. Similarly, the heat associated with the combustion of a substance under isobaric conditions is referred to as the *enthalpy of combustion. Enthalpy of formation* (ΔH_{form}) is the heat change when a chemical compound is formed from its component elements under isobaric conditions. The component elements must be in their most stable form, (for example, $C_{(graphite)}$, He, H_2, O_2). If the synthesis concerned is of 1 mol compound at pressure 1 atm and temperature 25°C, then the thermal product is referred to as the standard enthalpy of formation ΔH^0_{form} of the compound.

Consider an exothermic and isobaric reaction of the transformation of A to B:

$$A \rightarrow \quad B \; \Delta H = -153 \text{ kJ mol}^{-1}$$

The reverse reaction (conversion of B $\rightarrow$ A) will have exactly the opposite heat requirement:

$$B \rightarrow A \quad \Delta H = +153 \text{ kJ mol}^{-1}$$

In other words, the change in enthalpy of a reaction in one direction is equal and opposite to the change in enthalpy of the same reaction in the opposite direction. This law was first formulated in the 1780s by Lavoisier and Laplace, and bears the name of those two French chemists.

Hess, in the 1840s, discovered in addition that the change in enthalpy is independent of the pathway followed by the reaction from reagents to products. This means that the ΔH of a reaction that cannot be measured can be calculated as the sum of individual reactions of known enthalpy values.

As an example, let us calculate the ΔH^0_{form} of the reaction

$$6C_{(graphite)} + 6H_{2g} + 3O_{2g} \rightarrow C_6H_{12}O_{6s}$$

when

$$C_{(graphite)} + O_{2g} \rightarrow CO_{2g} \qquad\qquad \Delta H^0_1 = -393.1 \text{ kJ mol}^{-1}$$

$$H_{2g} + \tfrac{1}{2}O_{2g} \rightarrow H_2O_l \qquad\qquad \Delta H^0_2 = -285.5 \text{ kJ mol}^{-1}$$

$$C_6H_{12}O_{6s} + 6O_{2g} \rightarrow 6CO_{2g} + 6H_2O_l \qquad \Delta H^0_3 = -2821.5 \text{ kJ mol}^{-1}$$

The first two equations can be multiplied by 6 (that is, the same reaction occurring six times) and Lavoisier's law allows us to reverse the third:

$$6C_{(graphite)} + 6O_{2g} \rightarrow 6CO_{2g} \qquad\qquad \Delta H^0_1 = 6 \times (-393.1) \text{ kJ mol}^{-1}$$

$$6H_{2g} + 3O_{2g} \rightarrow 6H_2O_l \qquad\qquad \Delta H^0_2 = 6 \times (-285.5) \text{ kJ mol}^{-1}$$

$$6CO_{2g} + 6H_2O_l \rightarrow C_6H_{12}O_{6s} + 6O_{2g} \qquad \Delta H^0_3 = +2821.5 \text{ kJ mol}^{-1}$$

When adding the individual equations, we also add the enthalpies according to Hess' law:

$$\Delta H^0_{form} = 6\Delta H^0_1 + 6\Delta H^0_2 + \Delta H^0_3 = -1250.1 \text{ kJ mol}^{-1}$$

2.3 Entropy

The preceding paragraphs convey the idea that for a spontaneous transformation to occur, the final heat content must be smaller than the initial. For a thermally neutral reaction ($\Delta U = 0$), any increase in volume is thermodynamically undesirable because the enthalpy of the system will increase. Given that in enthalpic terms, a process H is favored by the

increase in mutual interactions (in example chemical bonds), every transformation in the universe must tend toward the formation of bonds. A
situation such as this would result in the condensation of everything, as
the only thermodynamically acceptable outcome would be the formation
of as many bonds as possible. Despite this, the material world generally
preserves *structure* and *form* that result from the balance of attractive and
repulsive forces. From the energetic point of view, this seems to violate
the principle of reduction in enthalpy. Clearly, to achieve equilibrium in a
given thermodynamic state (and to prevent the universe from shrinking
due to the formation of infinite bonds), an energetic entity is required that
can counteract the effect of enthalpy. What is this entity?

Let us consider a group of ideal gas molecules in a room, as per the left
part of Figure 2.1. Let us ignore attractive and repulsive forces between the
molecules and confine them in a small box, leaving all the remaining space
empty. Now imagine that the box suddenly disappears. The only movement that applies to the molecules is thermal motility, which is indirectly
measured by their temperature. This movement is random. It is obvious
that after the removal of the box, the molecules will not remain confined
in their previous volume. Clearly, each molecule will have moved away
from its starting position after a short time, the molecules spreading to
cover the new volume that is available to them, as per case B of Figure 2.1.
Why does this happen? Each molecule has probability p_1 of moving in a
particular direction and probability z of moving in any other direction
other than that denoted by p_1. A combined probability p can be defined for
the movement of a single molecule that will bring it into closer proximity
with another, stationary molecule. For two randomly moving molecules, p'
denotes the probability that they will approach one another and z' denotes
the probability that they will move farther apart. In this case $z' > p'$, meaning that the molecules are more likely to spread apart than to converge.
As the walls of the room are the only limit to their movement, the gas
will soon have expanded to occupy all the space available in the room. At
this point, all the molecules will be moving and maintaining the greatest possible average distance between themselves, which is equal to the
separation distance that they would have if they were static and evenly
distributed throughout the space. What has occurred from a phenomenological point of view? The molecules of the gas rearranged themselves
spontaneously in order to occupy the largest volume available to them, an
action that could be considered counter-enthalpic ($\Delta V \geq 0$). There is then a
tendency for a substance to move in such a way as to maximize the possible places it can occupy in the space available. The potential axes of movement can be described as the *degrees of freedom* of the movement of a body.
This tendency of systems toward the maximization of degrees of freedom
we call "entropy." It is apparent that the entropy S will be directly related

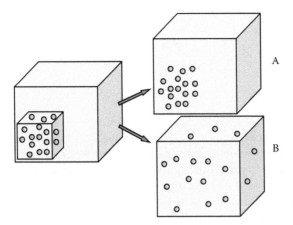

Figure 2.1 Graphical description of the experiment described in the text: Molecules initially confined in a restricted space will tend to spread so as to occupy the entire volume available to them (case B).

to the probability of a potential development of events P. The greater the probability of a development based only on spontaneous thermal motion of the molecules, the greater the entropy. The latter can be expressed quantitatively as proportional to the logarithm of every probability:

$$S = k \ln P \tag{2.13}$$

where k is the Boltzmann constant, for which the arithmetic value (R/N, the universal gas constant/Avogadro's constant) is explained later.

Let us return to the attempt to define energy from the first chapter. There, it was implied that "energy" can be defined as that which has the ability to drive a spontaneous change. In the preceding lines we said that the need for an increase in entropy drives systems to spontaneous change. Increasing the temperature accelerates the thermal motion of molecules and consequently facilitates the spontaneous transformation toward increased degrees of freedom. There is nothing about enthalpy in the above: There is no energy of mutual interactions, and neither is there any spontaneous reduction in volume. Consequently, entropy is related to energy that facilitates the moving apart of molecules from each other ($\Delta V \geq 0$), energy that increases with temperature.

Let us reconsider the case of the expansion of an ideal gas. For ideal gases, the internal energy U is independent of the volume. Because during the expansion $\Delta U = 0$, we have $q = -w$, that is, the heat gained by the system is equal to the work it carries out. Assuming that the process is reversible, we can write

$$-w = \int_{V_A}^{V_B} P dV = \int_{V_A}^{V_B} \frac{nRT}{V} dV = nRT \int_{V_A}^{V_B} \frac{dV}{V} \Rightarrow$$

$$-w = nRT \ln \frac{V_B}{V_A}$$

(2.14)

Let us consider again the previous example: M molecules are enclosed within a space A of volume V_A, which is itself enclosed in a larger space B of volume V_B. If we erase the walls of container A so that the molecules are free to move in the whole space B, then statistically we can say that the probability ρ_A that the molecules will be found in space A in relation to the probability ρ_B that they will be spread out into the whole space B after random movement is

$$\Delta S = \frac{-w}{T} = \frac{q}{T} = nR \ln \frac{V_B}{V_A}$$

(2.15a)

For 1 mol, namely N molecules (where N is Avogadro's number) and according to the above.

Let us denote the constant R/N by k. This is the Boltzmann constant that we met in the definition of entropy (see Equation (2.13)).

According to the above, if for example the volume V_B is ten times the volume V_A, and considering 1 mol of gas, the change in entropy during the expansion will be

$$\Delta S = S_B - S_A = R \ln (V_B/V_A) \approx 8.3 \ln 10 \approx 8.3 \times 2.3 = 19.1 \text{ J K}^{-1} \quad (2.15b)$$

This is a significant quantity of energy per unit temperature. Despite this, our imagination (which has provided the initial spark for many scientific discoveries) tells us that, although unlikely, it is not impossible for all the molecules to gather themselves back within the boundaries of space A at some point in time by pure chance (see case A in Figure 2.1). From the above mathematical relationships and for ΔS value of 19.1 JK^{-1}, the probability as calculated by 2.13 by inversing the logarithm (i.e., the probability that the molecules will be found within the limited space A rather than distributed throughout the larger volume B) approximates to e^{-23}. In other words, the probability of such an occurrence is vanishingly small.

The quantity of energy that relates to the entropy of a system at temperature T can be defined as the *entropic component* E_S.

$$E_s = -T\Delta S$$

(2.16)

The final form then of a material body or the energy of a state is determined by two factors: one that is based on bond formation (enthalpy) and one that is based on its free movement (entropic component[*]). Thus, every process that a substance undergoes (e.g., the formation of the tertiary structure of a protein) can be quantified energetically as the total of the two contributing factors. This final quantity of energy is called the *Gibbs free energy* ΔG and under isobaric conditions is described by the equation

$$\Delta G = \Delta H - T\Delta S \qquad (2.17)$$

Every population of molecules will, because of entropy, attempt to maximize its free movement throughout the greatest available volume. If we place in a glass one hundred "yellow" molecules and one hundred "black" molecules, then as long as these molecules do not mutually interact, they would soon be completely mixed. The presence of the other "colors" is of no consequence; each population simply claims all of the glass for itself. Where is entropy in all that? Let us assume now that the molecules are not free to move, but are bonded to other molecules by means of chemical bonds (one of enthalpy's ways of restricting molecular mobility). If these enthalpic interactions (bonds) are removed from the system, it will lose its form and disintegrate ("disorder"[†]).

2.4 Phase transitions

A *gas* is a material consisting of molecules that do not interact (or interact only weakly) with each other, moving freely (entropically) in all the space available to them due to thermal motion. If (enthalpic) bonds start forming between the molecules, their freedom of motion is restricted, the material becoming *liquid*. If even more bonds form, a lattice results, restricting the individual molecules in even more rigid positions in what is now a *solid*. We may perceive solids as entities with large enthalpy (due to the large number of bonds) and small entropic components (due to the restricted mobility of the individual molecules). As the entropic component TS is small in solids and large in gases, solids will be formed at low T, while gases will be formed at higher T, with the liquids somewhere in between them.

Let us consider the above in more detail, visualizing the heating up of a solid. As the temperature increases, increased molecular mobility results in the disruption and breaking up of the bonds between the molecules, which in turn results in a decrease in enthalpy (reduction in mutual

---

[*] Not "entropy"! The entropy S does not have dimensions of energy, while the entropic component TS does.

[†] The term "disorder" is not generally used in the current text. This is intentional, because entropy and disorder are clearly not identical. The term is used here solely to convey qualitatively the concept of the redetermination of the structure of a system after the lifting of enthalpic interactions. The disorder that is observed in this case is only a partial example of entropy.

interaction between molecules) and an increase in entropy (increase in the freedom of movement of the molecules). To approach the matter simply, we can distinguish two competing forces that act during the heating of a solid: (1) the forces that are exerted on the molecular lattice of the solid from the thermal motion of the molecules (which increase with increasing temperature and tend to break off molecules from the lattice), and (2) the forces that hold the molecules into the lattice (chemical bonds, other interactions). At low temperatures, the forces maintaining the solid structure (2) are stronger, keeping the molecules in their place. As the temperature rises, the forces (1) become stronger. At a particular temperature, the forces of thermal movement of the individual molecules (1) will equal and begin to exceed the forces that hold the molecules in place (2), with the result that the molecules will abandon their places in the solid lattice and move with relative freedom within the confining space. This phenomenon when occurring en masse is referred to as *melting*, and the temperature at which it occurs is the *melting point* T_{melt}. The opposite course, that is, the lowering of the temperature from $T > T_{melt}$ to $T = T_{melt}$ and continuing to still lower temperatures leads to solidification of the material. Solidifying starts at a point (the freezing or crystallization point) at which the forces of thermal motion become weaker than the forces of attraction between molecules. Lowering the temperature reduces the extent of thermal motion. At the freezing point, the forces (1) and (2) are equal to each other. At this point, the liquid becomes a solid; that is to say, its molecules lose their motility and become locked in clearly defined positions.

If a liquid is heated further to a temperature $T_{boil_{BP}}$ (the boiling point), the interactive forces that provide some consistency to the liquid (later we shall see how these relate to the viscosity) are finally overcome by the thermal motility of the molecules and the liquid makes the transition to gaseous phase, that is, to a state in which the molecules interact only minimally with each other and move freely throughout the space available. The reverse determines the condensation point, at which the attractive forces between molecules limit their motility, converting the gas to a liquid.

Let us now observe the same phenomenon by studying energies rather than forces. As discussed, the entropy S of a system relates to the free movement of the molecules that comprise it: High freedom of movement corresponds to high entropy values while low freedom of movement corresponds to low entropy values.

At every temperature, the relationship $\Delta G = \Delta H - T\Delta S$ can be used to determine the most stable state of a system, namely that with the lowest Gibbs free energy. At low temperatures (at which solids are generally found), the enthalpy H is generally large because forces (bonds) are applied between the molecules that immobilize them in their positions. On the other hand, the entropic factor TS of a solid system is small as the temperature is low (small T) and the immobilized molecules have limited freedom of movement (small S). At the melting point, under reversible conditions, the heat q_{rev} supplied to the system is not used to increase the temperature but rather to break the bonds between the molecules. This value correlates to the enthalpy of fusion ΔH_{fus}, or *latent heat of fusion* as it is more usually known. As bonds are destroyed, the system moves rapidly toward a state that maximizes its entropy. Because the system is at constant temperature and (let us assume) constant pressure, the entropy that relates to the melting ΔS_{fus} is defined as

$$q_{rev} = T_{fus}\Delta S_{fus} \Rightarrow \Delta S_{fus} = \frac{q_{rev}}{T_{fus}} = \frac{\Delta H_{fus}}{T_{fus}} \tag{2.18a}$$

Once melting is complete, the temperature rises again and ascends the region between the melting point and the boiling point (the range in which the system is in a liquid state). Within this range, the values of H are smaller than those in the solid state (as the bond-associated interactions between the molecules are weaker) and the values of S are larger (because the molecules have greater freedom of movement). The change in entropy for a narrow temperature range (here for the transition from T_1 to T_2) can be given by

$$\Delta S_1 = C_p \ln \frac{T_2}{T_1} \tag{2.18b}$$

One must bear in mind that the heat capacity C_p changes with temperature; but in practice and under restrictions, it may be considered constant for very small areas of temperature.

In the liquid phase, the supplied heat is stored in the system, increasing the mobility of the individual molecules, thus raising the temperature T. The change in temperature ΔT for every unit of heat applied is determined, in our case, by the thermal capacity of the material in liquid phase under constant pressure C_p.

When the temperature reaches the boiling point, we can determine the entropy of vaporization ΔS_{vap} and the enthalpy of vaporization ΔH_{vap} for the transition from the liquid to gaseous state (always assuming that the change is reversible and isobaric):

$$\Delta S_{vap} = \frac{q_{rev}}{T_{vap}} = \frac{\Delta H_{vap}}{T_{vap}} \tag{2.19}$$

At this temperature, the heat supplied to the liquid breaks down the bonds that retain the structure of the liquid. Quantitatively, this presents itself as a reduction in enthalpy (latent heat of vaporization, ΔH_{vap}). The reduction in enthalpy leads to an increase in the entropic factor *TS*. More particularly, the increase in entropy means that the molecules will claim a much larger space, spreading to fill all of the volume available to them.

In the gas phase, the supplied heat raises the temperature *T*. The change in temperature ΔT is determined by the thermal capacity of the material in the gas phase under constant pressure C_p.

The above description constitutes the basis of phase diagrams. These diagrams present the individual phases of a material as a function of the variables of state (*P*, *V*, *T*). In Figure 2.1, the phase diagram of a substance is presented schematically (not to scale). The sublimation line A gives the vapor pressure of the solid phase and, for a given pressure *P*, the sublimation point (the temperature for the direct transition from solid to gas phase, e.g., from ice to steam for water). The vapor pressure line B gives the boiling point (the temperature for the transition from liquid to gas phase, e.g., from liquid water to steam) for a given pressure *P*. The melting line C gives, for every value of pressure, the melting temperature (the temperature for the transition from solid to liquid phase, e.g., from ice to liquid water). The three plot lines meet at the *triple point*. This is characteristic for every material and indicates the point at which the three phases (liquid, solid, and gas) can coexist. For water, the triple point is ~273.16 K and ~4.59 mmHg. The vapor pressure line ends at the critical point, beyond which the boundary of the two phases (liquid and gas) ceases to exist. Here we define an additional fourth phase, the *supercritical state*, which has special properties. The pressure and temperature that equate to the critical point are referred to as the *critical pressure* and *critical temperature*, respectively. The critical point is, like the triple point, characteristic for every substance.

In the preceding paragraphs, we referred to the way in which the entropy determines the spontaneous character of a reaction and the way in which (under isobaric conditions) the result of the subtraction $\Delta H - T\Delta S$ determines the equilibrium point. It must never be forgotten that the equation $\Delta G = \Delta H - T\Delta S$ applies *only* under conditions of constant pressure (isobaric conditions). Such conditions prevail at the Earth's surface (*P* = 1 atm), and apply if the reactions or processes of interest take place in an open container. There are, however, applications (e.g., in the case of canning or vacuum-packaging) where the pressure is altered. In this case, we must begin whichever calculations we wish to perform with the starting relationship

$$G = U + PV - TS \tag{2.20}$$

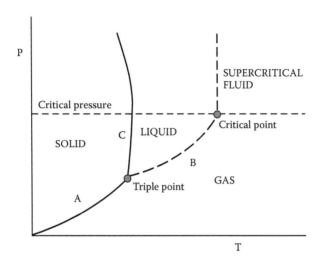

Figure 2.2 Typical phase diagram of a material. The regions of pressure and temperature in which it is solid, liquid, gas, or supercritical fluid are delineated. Note the triple and critical points at which three states coexist and the phase boundaries (lines) at which two states coexist.

and the first differential of this:

$$dG = dU + PdV + VdP - TdS - SdT \tag{2.21}$$

From the basic definition of free energy change as the sum of changes in heat and work, and assuming reversible processes, we have

$$dU = dq + dw \tag{2.22}$$

From the definition of entropy $dq = TdS$ and of isobaric work $dw = -PdV$ derives:

$$dU = TdS - PdV \tag{2.23}$$

Then the differential of dG becomes

$$dG = VdP - SdT \tag{2.24}$$

This relationship is very useful for systems of constant volume and changing pressure and temperature as it provides information for the way in which the free energy (and consequently the equilibrium) changes with the pressure and temperature. Consider two phases of a substance, for example a liquid and its vapor. If the temperature is altered by dT and the pressure by

dP, the relationship $dG = VdP - SdT$ that we saw previously becomes for the liquid (l)

$$dG_1 = V_1 dP - S_1 dT \tag{2.25}$$

and for the gas (g)

$$dG_g = V_g dP - S_g dT \tag{2.26}$$

On balance, the Gibbs free energy is the same in the solid and the liquid. Consequently,

$$V_1 dP - S_1 dT = V_g dP - S_g dT \tag{2.27}$$

$$\frac{dP}{dT} = \frac{S_g - S_l}{V_g - V_l} \tag{2.28}$$

Defining the entropy of vaporization $\Delta S_{vap} = S_g - S_l$ and the equivalent volume $\Delta V_{vap} = V_g - V_l$ and accounting for the fact that we have equilibrium ($\Delta G = 0 => \Delta H_{vap} = T\Delta S_{vap}$),

$$\frac{dP}{dT} = \frac{\Delta S_{vap}}{\Delta V_{vap}} = \frac{\Delta H_{vap}}{T\Delta V_{vap}} \tag{2.29}$$

The differential dP/dT at equilibrium will be the formula for the latent heat of the phase change ΔH:

$$\left(\frac{dP}{dT}\right)_{equil} = \frac{\Delta H}{T\Delta V} \tag{2.30}$$

The mathematical relationship above is known as the *Clapeyron equation* and is particularly useful because, with this, we can calculate the changes that take place during phase transitions (vaporization, sublimation, melting).

In vaporization, considering that the volume of a gas is considerably greater than that of the liquid from which it was formed, we can take $\Delta V_{vap} \approx V_g$. In addition, considering ideal gases, the volume of the produced gas can be rewritten as RT/P. Hence, we can write

$$\frac{dP}{dT} = \frac{P\Delta H_{vap}}{RT^2} \Rightarrow \frac{dP}{P} = \frac{\Delta H_{vap} dT}{RT^2} \Rightarrow \int \frac{dP}{P} = \int \frac{\Delta H_{vap} dT}{RT^2}$$

$$\Rightarrow \ln P = -\frac{\Delta H_{vap}}{RT} + K \tag{2.31}$$

The expression in Equation (2.31) is called the Clausius–Clapeyron equation. This equation, while approximate, is especially useful because it allows the calculation of the latent heat of vaporization of a substance based on a given vapor pressure and temperature.

2.5 Crystallization

A basic characteristic of solids and many liquids is their property of forming clearly defined structures, which are called *crystals*. Crystals are formed when the structural components of a material (i.e., the ions or molecules that comprise it) place themselves in particular positions close to one another in order to achieve the minimum possible free energy by means of the formation of large numbers of bonds between them (minimization of enthalpy). The overall structure defined by the ordered positioning of the components is called the *crystal lattice*. Because crystallization entails the stabilization of the molecules in very particular positions, the organized molecules are deprived of a significant degree of freedom, and hence the entropy is reduced. For this reason, the formation of crystals tends to be favored by low temperatures at which the entropic contribution TS is small.

2.6 Application of phase transitions: Melting, solidifying, and crystallization of fats

Fats and oils (in food technology, they are the same substances in their solid and liquid forms, respectively) are primarily composed of triglycerides and different derivatives thereof. While they clearly contain compounds of other categories in smaller proportions, for teaching purposes we can at present safely regard a fat or an oil as a collection of triglycerides.

Triglycerides are triacyl esters of glycerol, that is, a molecule of glycerol esterified at its three hydroxyl groups with three fatty acids (see Figure 2.3). In practice, a free triglyceride will arrange itself in space approximately as presented on the left of the figure, with the molecule of glycerol at the center and the fatty acid chains spread out far from each other in order to maximize the degree of freedom and hence the entropy of the system.

Two neighboring fatty acid chains can develop hydrophobic interactions between their methyl groups ($-CH_2-$), leading to a reduction in both enthalpy and entropy of the whole system. The nature of these interactions is discussed in a later chapter. These interactions are easier at lower temperatures when the thermal motility of the molecules is limited and the energy of the bonds is sufficient to immobilize them in their respective places in a crystalline lattice. The hydrophobic attractions between

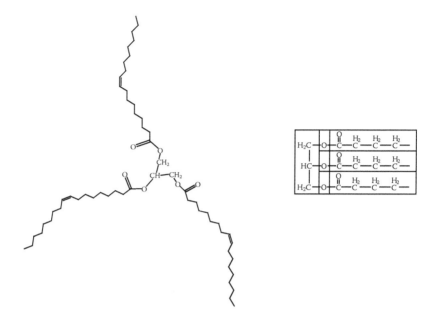

Figure 2.3 Schematic representation of a free triglyceride consisting of three monounsaturated fatty acids.

two neighboring chains are clearly stronger when there are many points of interaction—that is, many methyl groups. Consequently, fatty acids with large carbon chains tend to form stronger bonds, with the result that their triglycerides melt at higher temperatures than triglycerides with lower-molecular-weight fatty acids.

Triglycerides display *polymorphism* (from the Greek words *poly*—many and *morphe*—shape), meaning that they can be found with more than one crystalline structure, depending on the temperature and the cooling conditions (principally the rate of change of temperature).

Highly schematic illustrations of typical crystalline structural arrangements of triglycerides are presented in Figure 2.4. For the convenience of the reader, the fatty acids appear as tiles that intersect the plane.

Because every fat and oil has a different composition regarding the fatty acids that its triglycerides contain (and their relative positions on the molecule), they crystallize into different forms. These are also dependent on the conditions of heating or cooling. The α-form or polymorph has the lowest melting point. In this form, the ends of the fatty acids have a relatively relaxed configuration between them. It is usual for fats to crystallize into the α-form during rapid cooling. These polymorphs tend to revert with time into denser—and thus stabler—forms. Such is the β′-form, which can retain its structure for some time, depending on the chemical composition of the fat. Such structures can eventually transform

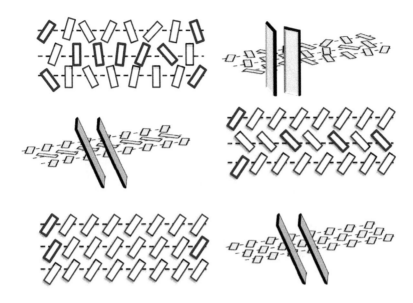

Figure 2.4 Highly schematic depiction of some forms of polymorphism in the arrangement of triglycerides in their crystal lattice in a fat. Vertical and side views of different crystallization forms are juxtaposed.

into the denser β-form, where the carbon chains are arranged parallel to one another. The changes are *monotropic* (irreversible), progressing from the unstable to the stable forms, and are not the same for all fats; that is, some polymorphs may not form at all, depending on the conditions and the chemical composition. As the denser forms are more stable, and thus involve stronger interactions, α-polymorphs tend to have a lower melting point, followed by the β′ and β polymorphs. These two normally have higher melting points.

From Figure 2.3 it is apparent that the basic prerequisite for the creation of a strong crystalline lattice is good contact from one fatty acid chain to the neighboring chain. As single bonds allow the free rotation of carbon atoms around their axis, a chain that exclusively contains single bonds between the carbon atoms can adopt a straight configuration (Figure 2.4). The existence of double bonds impedes this straightening, as rotation around the axis of a double bond is not possible. In addition, in fatty acids, the double bonds are generally in the *cis* configuration, that is, the large chains are located on the same side of the bond. Thus, the presence of double bonds (i.e., unsaturated fatty acids) forces the chain to fold, as shown in the figure. A folded chain is difficult to accommodate in a crystal lattice—this can happen eventually but will require a lower temperature. As a result, triglycerides with many unsaturated fatty acids (i.e., fatty acids with many double bonds) tend to be liquid, while at the

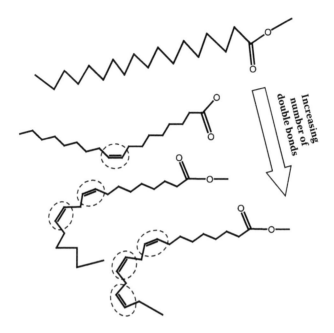

Increasing number of double bonds

Figure 2.5 Chain curvature under the influence of multiple *cis* double bonds. Observe how much more difficult it is for polyunsaturated fatty acids to position themselves next to one another, and also consider how this influences their physical state (solid or liquid).

same temperature the equivalent saturated triglycerides are solid. For this reason, hydrogenated fats tend to be solid despite the fact that they may originate from liquid oils.

As the temperature of a solidified fat rises, the entropic component $T\Delta S$ rises proportionally. Seeking greater freedom of movement, the molecules move more vigorously. When the kinetic energy becomes greater than the energy that is stored in the bonds between the fatty acids, the triglycerides detach and the system becomes liquid. If the fat contains triglycerides with different fatty acids, some crystals will melt and others will not. This phenomenon will be familiar to whomever has placed some oil in the refrigerator and observed its transformation into a colloidal suspension of fat in oil, which with further chilling solidifies completely.

2.6.1 Chocolate: The example of cocoa butter

Chocolate melts in the mouth but not in the hand.[*] The key to the very particular melting and crystallization of chocolate lies in the fact

---

[*] According to Coultate (1999).

that cocoa butter has six forms, with melting points ranging from 17.3°C to 36.4°C. Of these six forms, the fifth (β-3, m.p. 33.8°C) satisfies the requirement of a material that melts in the mouth but not in the hand. The chocolate maker must regulate the crystallization with fractional chilling/reheating just before the required temperature in order for the product to crystallize while avoiding undesired crystal types. Obviously, the art of the chocolate maker, cultured for centuries (millennia in Mexico) before our time, is much further ahead of the corresponding work of the chemist and food technologist. This provides ample proof that cooking and chocolate making are among the oldest forms of chemistry that flourished long before physics and chemistry combined to transform chocolate making into a physical science!

2.7   Chemical potential

We have seen that the basic condition to bring about equilibrium to a process or reaction is for further change in Gibbs free energy to be zero ($\Delta G = 0$). It is possible for two or three groups of different components to exist within a system. If all the molecules are evenly distributed throughout the container, then for each of them there is the possibility of a decrease in enthalpy via the formation of bonds; that is, each of them can contribute to a further reduction in the free energy G of the system.

If a system contains N substances instead of one, the Gibbs free energy will include the sum of the molecular concentrations of each substance n_1, $n_2, n_3, ..., n_N$.

$$dG = VdP - SdT + \left(\frac{\partial G}{\partial n_1}\right)_{T,P,n_j} dn_1 + \cdots +$$

$$\left(\frac{\partial G}{\partial n_i}\right)_{T,P,n_j} dn_i + \cdots + \left(\frac{\partial G}{\partial n_N}\right)_{T,P,n_j} dn_N \Rightarrow \qquad (2.32)$$

$$dG = VdP - SdT + \sum_{i=1}^{N} \mu_i dn_i$$

The factor μ_i is called the *chemical potential* of substance i and is a measure of the potential, or lack thereof, of the substance in question to react with other molecules in the system under study. The chemical potential equates to the change in Gibbs free energy of the system that is caused by the addition of the substance concerned. The definition of chemical potential is

$$\mu_i = \left(\frac{\partial G}{\partial n_i}\right)_{T,P,n_j} \qquad (2.33)$$

where j refers to every substance other than i.

Consider a system that contains, among N substances, the substance i. If we transfer molecules of i from point A to point B, then

$$dG = [\mu_i(B) - \mu_i(A)]dn_i \qquad (2.34)$$

To achieve equilibrium for a small number n_i, chemical potentials must be equal. $\mu_i(A)$ must apply. That is to say, a process arrives at equilibrium when all individual similar molecules have obtained the same chemical potential. This is a very useful definition of the equilibrium point.

EXERCISES

2.1 The energy for the hydrogenation of ethylene at 25°C is −32.6 kcal mol^{-1} while the heat of combustion for ethane, n-butane and hydrogen are −372.8, −688.0, and −68.32 kcal mol^{-1}, respectively. Calculate the heat for the following n-butane pyrolysis: $C_4H_{10g} \rightarrow 2C_2H_{4g} + H_{2g}$ at 25°C.

Solution: Write the equation for the ethylene hydrogenation and the combustions: $C_2H_{4g} + H_{2g} \rightarrow C_2H_{6g}$; $C_2H_6 + \frac{7}{2}O_{2g} \rightarrow 2CO_2 + 3H_2O_l$; $C_4H_{10g} + \frac{13}{2}O_{2g} \rightarrow 4CO_{2g} + 5H_2O_l$; $H_{2g} + \frac{1}{2}O_2 \rightarrow H_2O_l$; rearrange and multiply the equations so as to sum $C_4H_{10g} \rightarrow 2C_2H_{4g} + H_{2g}$. Add the components. According to Hess' law, the sum of the enthalpies gives the answer.

2.2 Calculate the entropy for the melting of 25 g of ice, initially stored at −15°C, then brought to 25°C. Consider the specific heat of melting for water to be 79.7 cal g^{-1}, the heat capacity of ice $C_{H_2Os} = 0.5$ cal g^{-1}, and that of liquid water $C_{H_2O_l} = 1.0$ cal g^{-1}.

Solution: Apply Equation (2.18b) for the heating up to 0°C; then Equation (2.18a) for 0°C; then Equation (2.18b) again for the heating up to 25°C. Add the entropies. Be careful with the units!

2.3 Calculate the standard entropy of chlorine in its gaseous form at 25°C, taking into consideration the following specific free energies and enthalpies of formation and specific entropies:

$\Delta G^\circ_{f,NaCl(s)} = -91.7$ kcal mol^{-1}; $\Delta H^\circ_{f,NaCl(s)} = -97.8$ kcal mol^{-1}; $S^\circ_{Na(s)} = 12.2$ cal mol^{-1} K^{-1}; $S^\circ_{NaCl(s)} = 17.3$ cal mol^{-1} K^{-1}.

Solution: Find $\Delta S^\circ_{f,NaCl(s)}$ from Equation (2.17); $\Delta S^\circ_{f,NaCl(s)}$ should be equal to the differences between the sum of the products and the reagents' entropies for the equation $Na_{(s)} + \frac{1}{2}Cl_{2(g)} \rightarrow NaCl_{(s)}$. Solve for $S^\circ_{Cl(g)}$.

2.4 A substance evaporates at 353.3 K. Considering that its latent heat of vaporization is 30.7 kJ mol^{-1}, calculate the change in entropy during the evaporation of 5 mol of this substance.

Solution: Use Equation (2.19); multiply the result by 5.

chapter three

The thermodynamics of solutions

3.1 From ideal gases to ideal solutions

It is safe to consider, in very broad and generic terms, that a gas is a group of molecules that move independently of each other, and thus entropically spread out so as to occupy any volume available to them. This motility is mostly thermal motion, the extent of which is reflected in the system's temperature. Interactions can occur between individual gaseous molecules but, with notable exceptions, they are relatively weak. As the temperature is reduced, the intermolecular interactions tend to be stronger, as bonds form in order to accommodate the reduction in mobility. This *liquid* system has two new characteristics as compared to the gas: (1) the newly-formed bonds link together a series of molecules in a way that a force applied to any one of them can be distributed to neighboring molecules, eventually giving rise to *viscosity*, which is discussed in a subsequent chapter; and (2) the new bonds are strong enough to pin the molecules close to their neighbors; they cannot expand to occupy any volume available to them.

The discussion in the previous chapters concerned gaseous systems, but in food science the reactions take place primarily in liquid solutions. It is thus necessary to give some attention to the peculiarities of the thermodynamic behavior of solutions. As mentioned in the previous paragraph, the motion of a liquid molecule is restricted by means of bonds with other molecules in its proximity. Such molecules cannot be readily released from their liquid matrix to become gaseous. In this respect, the transition from liquid to gas for a group of molecules is an interplay between the reduction in enthalpy ΔH due to the formation of bonds (which favors the liquid state) and the increase in entropic contribution $T\Delta S$ due to the increase in temperature (which favors maximization of entropy, thus spreading the molecules in their gas form). As temperature increases, so does $T\Delta S$, driving the system toward volatilization.

Let us consider a liquid in an open container at temperature T and pressure P. At a given temperature, even below the boiling point, a number of molecules are expected to leave the liquid state and partition as gas. The amount of such molecules depends on the interplay between ΔH and $T\Delta S$, with high T favoring the partition into the gas phase. The pressure needed to avoid the volatilization of these molecules is called the *vapor pressure* of the system.

The basis of the examination of the physicochemical properties of solutions is *Raoult's law*, according to which the vapor pressure p of a substance that exists in solution is equal to

$$p = p^* X \tag{3.1}$$

where p^* is the vapor pressure of the isolated substance and X its mole fraction, which is defined as

$$X = \text{(Moles of the substance)/(Total moles of the solution)}$$

For a mixture of two substances (A and B), Raoult's law can be written as

$$p = p_A^* X_A + p_B^* X_B \tag{3.2}$$

where p_A^* and p_B^* are the vapor pressures of the isolated substances A and B, respectively, while X_A and X_B are the respective mole fractions in their liquid mixture. Analogous to ideal gases, an *ideal solution* is a liquid system that obeys Raoult's law.

Now consider a system at constant temperature, where the dT component of Equation (2.24) is zero. Let us examine the dependence of the free energy G on the pressure P and the temperature T (because they play a role in Equation (2.24)). At constant temperature, the equation $dG = VdP - SdT$ gives

$$dG = VdP \Rightarrow \int_{G_1}^{G_2} dG = \int_{P_1}^{P_2} VdP = nRT \int_{P_1}^{P_2} \frac{dP}{P} = nRT \ln \frac{P_2}{P_1} \tag{3.3}$$

That is to say,

$$G_2 - G_1 = \Delta G = nRT \ln \frac{P_2}{P_1} \tag{3.4}$$

Consider the reference pressure P^o of a component equal to 1 atmosphere (atm) and G^o the free energy at 1 atm; then Equation (3.4) can be written for any pressure P and free energy G as

$$G - G^o = nRT \ln \frac{P}{P^o}$$

$$\Rightarrow G_{vap} - G_{vap}^o = nRT \ln \frac{p}{p^o} \tag{3.5}$$

The last part of the equation substitutes pressure for the equivalent vapor pressure p of a vapor separating from a solvent; here, p_o is a vapor pressure equal to 1 atm, while G_{vap} and G^o_{vap} are the equivalent free energy values of the vapors.

Considering the definition of vapor pressure, it is safe to assume for the gaseous phase that the partial pressure of a component is equal to its vapor pressure.

Let us consider the partial evaporation of the molecules of a solvent equilibrating at partial pressure p_{vap}. Because equilibrium exists, ΔG for the transition between liquid and gas will be zero. Let us consider the compression of this vapor from an initial pressure p_{vap} (equal to p_{vap}^* for a single-component system, $X_{vap} = 1$) to 1 atm. According to Equation (3.5),

$$G^o - G = -nRT \ln \frac{p}{p^o} = -nRT \ln \frac{p}{1\ atm} = -nRT \ln p_{dim}^* \qquad (3.6)$$

remember that p_{dim}^* is dimensionless.

For the direct evaporation of a liquid at 1 atm to vapor at 1 atm, ΔG can be calculated by summing the free energies associated with the evaporation of liquid to a specific vapor pressure ($\Delta G = 0$), plus the free energy for the compression of the vapor to 1 atm (Equation (3.6)); for 1 mol: that is,

$$\Delta G^o = -RT \ln p_{dim}^* \qquad (3.7)$$

$$G^o_{vap} = G^o_{liq} - RT \ln p_{dim}^* \qquad (3.8)$$

Because the non-zero part is a conversion of 1 mol in 1 atm, the overall conversion of 1 mol at 1 atm from liquid to vapor is $\Delta G = \Delta G^o = G^o_{vap} - G^o_{liq}$.

Assuming constant pressure, at any conditions where equilibrium exists (that is, $G_{vap} = G_{liq}$), Equation (3.5) gives for 1 mol and the dimensionless p_{dim}:

$$G_{vap} - G^o_{vap} = G_{liq} - G^o_{vap} = RT \ln \frac{p}{p_o} \qquad (3.9)$$

Hence,

$$G_{liq} = G^o_{vap} + RT \ln \frac{p}{p_o} = G^o_{vap} + RT \ln p_{dim} \qquad (3.10)$$

again taking the liberty to ignore $p^o = 1$ atm and changing into the dimensionless p_{dim}.

Substituting into Equation (3.8) into (3.10), we obtain

$$G_{liq} = G_{liq}^0 - RT \ln p^* + RT \ln p \tag{3.11}$$

According to Raoult's law, the partial pressure p of a substance i contained in an ideal mixture at a molar ratio X_i is defined as

$$RT \ln p = RT \ln p^* X_i = RT \ln {}^*X_i + RT \ln p^* \tag{3.12}$$

where X_i is the mole fraction of substance i. That is, the total pressure is the sum of the partial pressures of the individual components of the system.

For an isolated system, the Equation (3.12) can be written in terms of free energy:

$$G_{liq} = G_{liq}^0 - RT \ln p^* + RT \ln p^* + RT \ln X_i \tag{3.13}$$

$$G_{liq} = G_{liq}^0 + RT \ln X_i \tag{3.14}$$

This is a very important equation, as it can provide the free energy of a solvent on the basis of its molar fraction in a liquid mixture, such as when it contains a solute. Because it is based on the assumptions of ideal gases, it can only describe ideal solutions; however, it is very useful in helping to understand the basic thermodynamic concepts of binary liquid mixtures.

3.2   Fractional distillation

Now consider a two-component system comprised of substances A and B. According to Raoult's law, for the equilibrium vapor composition $X_A(g)$ and $X_B(g)$ and liquid phase composition $X_A(l)$ and $X_B(l)$ we have for A and B, respectively, that

$$X_A(g) = \frac{p_A}{p} = \frac{X_A(l)p_A^*}{p} \tag{3.15}$$

and

$$X_B(g) = \frac{p_B}{p} = \frac{X_B(l)p_B^*}{p} \tag{3.16}$$

Here, p is the total pressure ($p = p_A + p_B$).

Dividing the equations gives us

$$\frac{X_A(g)}{X_B(g)} = \frac{X_A(l)}{X_B(l)}\frac{p_A^*}{p_B^*} \tag{3.17}$$

This means that the vapor will be richer in the more volatile component in comparison with the liquid. The transfer of vapor out of the liquid brings about the enrichment of the vapor with the more volatile ingredients and the enrichment of the liquid mixture with the less volatile ingredients. The difference in composition between vapor and liquid allows us to separate the substances by means of distillation.

In Figure 3.1, the vapor pressure of a solution is presented in relation to the composition of the mixture. The upper (dashed) line represents the vapor pressure for every composition of the *solution*. The lower (solid) line gives the vapor pressure for every composition of the *vapor*. If we bring to a boil a mixture of two components A and B with an initial composition of $X_A(l)$ and $X_B(l) = 1-X_A(l)$, then the initial vapor composition will be $X_A(g)$ and $X_B(g)$, which, according to the figure, is richer in component B. If we isolate, condense, and reboil the vapor phase, the resulting vapor would be richer still in component B, with composition $X_B'(g)$.

This is the principle of fractional distillation: a simple one-stage distillation is one stage on the graphic, with a product of content $X_B(g)$. The second stage ("second distillation column") is the stage during which,

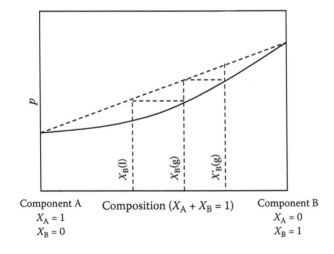

Component A   Composition ($X_A + X_B = 1$)   Component B
$X_A = 1$ $X_A = 0$
$X_B = 0$ $X_B = 1$

Figure 3.1 Change in vapor pressure of a mixture of two substances, A and B. The lines represent the vapor pressure for every composition of the *solution* (dashed line) and for every composition of the *vapor* (solid line).

upon second boiling, the concentration of B increases from $X_B(g)$ to $X_B'(g)$. In alcoholic beverage production, sequential distillations are carried out in order to increase the ethanol content of the fermented product from values on the order of 10% to 20% to values on the order of 40% to 50% (ethanol is more volatile than water).

In real two-component liquid systems, it is possible to have a clear minimum or maximum on the boiling point–composition plot. In the case of Figure 3.2, as long as the composition of the system is found to the left of the maximum point, mixtures of the two substances can be isolated by

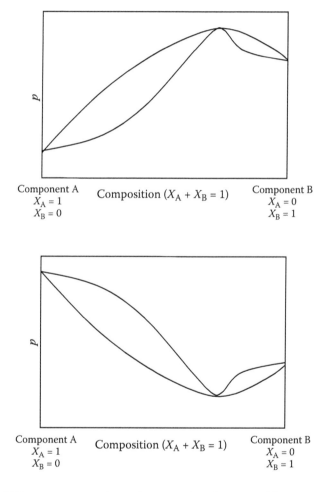

Figure 3.2 Vapor pressure plots of azeotropic mixtures with two components (A and B) that show an azeotropic maximum (above) and an azeotropic minimum (below).

means of fractional distillation, but the pure component B (and whatever mixture with a composition beyond the maximum) cannot. Such mixtures are called *azeotropic* and can create complications for the distillation of pure components.

The problems created by azeotropic mixtures can be countered with a series of techniques such as changes in pressure or the addition of a third material to the system. In some circumstances, the presence of azeotropic mixtures may be desirable, as they allow mixtures of a clearly defined concentration to be obtained. For example, hydrogen chloride (which boils at −80°C at 1.013 bar) and water form an azeotropic mixture at 108.584°C with a hydrochloric acid content of 20.22% by weight. This is an elegant way to produce aqueous solutions of hydrochloric acid of this concentration.

In the case that strong bonds (such as hydrogen bonds) are formed between the components of a mixture, it is possible for azeotropic minima to occur (see Figure 3.2, bottom). The vapor pressure of such systems tends to be lower than would be predicted by Raoult's law on account of the mutual interactions between the molecules. For example, in the case of alcohol and water mixtures that are familiar to the drinks industry, the alcohol (which boils at 78.3°C at 1.013 bar) and the water form an azeotropic minimum that boils at 78.174°C and contains approximately 4% water by weight.

3.3   Chemical equilibrium

The Second Law of Thermodynamics that was presented previously can be applied in order to predict in which direction a process will proceed and to which point this will occur. This process is called *reversible* when the system can be changed from state A to state B and subsequently return to A. Let us consider a gas contained within a closed space and in contact with a piston, with the pressure on both sides of the piston being the same. If the external pressure decreases, the gas will expand. Vice versa, if the external pressure increases, the gas will be compressed. If these changes in pressure are very small and take place very slowly (in order to bring about equilibrium between each initial and final state), then these processes are, in practice, reversible. A process can be defined as *reversible* only if it is the product of a series of infinitely small sequential changes from one state to another. The changes must occur extremely slowly, so that each begins from a state of equilibrium. In general terms, whatever does not conform to the above is considered an *irreversible* transformation. A characteristic of the latter is that it occurs spontaneously. Typical examples are the flow of heat from a warm point to a cold point, or the explosive expansion of gases in the case of an explosion.

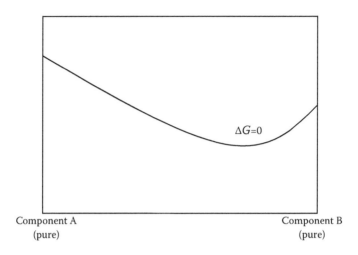

Component A   Component B
(pure)   (pure)

Figure 3.3 Change in free energy dG during the transformation of a substance A to B. Note that the reaction comes to equilibrium without the whole quantity of A having converted to B.

Let us consider a process in which the molecules of a substance A are transformed into the molecules of a substance B. This reaction would be irreversible, that is, would be spontaneous, if

$$dG < 0 \Rightarrow dH - TdS < 0 \tag{3.18}$$

and

$$dS > \frac{dH}{T} \tag{3.19}$$

The reaction will proceed for as long as this condition is satisfied. When

$$dH - TdS = 0 \tag{3.20}$$

—that is, when the free energy ΔG of the reaction becomes equal to zero—the reaction will stop. Thus, the reaction $\alpha A \rightarrow \beta B$, in which α mol of A react to form β mol of B, will proceed until the point at which the total free energy of the system equals zero (Figure 3.3).

Assuming that A and B are gases that react in a finite volume V, then $dG = VdP - SdT$ applies. If the temperature remains constant for the duration of the reaction ($dT = 0$), we have $dG = VdP$. In the previous paragraphs, we saw that in this case we can say for component A that

$$G_A - G_A^o = \alpha RT \ln \frac{P_A}{P_A^o} \tag{3.21}$$

Similarly, for component B:

$$G_B - G_B^o = \beta RT \ln \frac{P_B}{P_B^o} \tag{3.22}$$

Considering a reaction such as $\alpha A \rightarrow \beta B$ instead of single-mole processes and bringing together the components, we end up with the following relationship, assuming that $\Delta G^o = G_{products} - G_{reactants}$:

$$-\Delta G^o = RT \ln \frac{\left(P_B/P_B^o\right)^{\beta}}{\left(P_A/P_A^o\right)^{a}} \tag{3.23}$$

in which ΔG^o is the difference in Gibbs free energy between the reactants and the products under standard conditions (e.g., for gases, this is generally 1 atm). The ΔG^o concerns the "full" hypothetical reaction of reactants $\rightarrow$ products and is not related to the equilibrium. ΔG, on the other hand, refers to the energetic difference (products – reagents) at *every point* in the reaction, obviously including the ΔG of the equilibrium. It is clear then that while ΔG can take many values according to the conditions, for every reaction, only one ΔG^o can be defined per temperature.

The fraction in the logarithm in Equation (3.23) contains the partial pressures of the reactants at the start of the reaction (P^o) and their partial pressures in the final state at equilibrium (P). Let us retain the fact that the reaction $\alpha A \rightarrow \beta B$ does not presuppose the complete disappearance of the reactants, as such an occurrence would make the denominator equal zero. As is apparent from Figure 2.5, at an intermediate state, ΔG becomes zero and equilibrium occurs. In the case of gases, and specifically in the reaction $\alpha A \rightarrow \beta B$, the ratio

$$\frac{\left(P_B/P_B^o\right)^{\beta}}{\left(P_A/P_A^o\right)^{a}}$$

is defined as the equilibrium constant of the reaction K_P. The above relationship may be rewritten as

$$-\Delta G^o = RT \ln K_p \tag{3.24}$$

For a hypothetical reaction,

$$\alpha A + \beta B + \gamma C + \ldots \rightarrow xX + yY + zZ + \ldots$$

The constant K_P is given by the relationship:

$$K_P = \frac{\left(P_X/P_X^\circ\right)^x \left(P_Y/P_Y^\circ\right)^y \left(P_Z/P_Z^\circ\right)^z \cdots}{\left(P_A/P_A^\circ\right)^\alpha \left(P_B/P_B^\circ\right)^\beta \left(P_\Gamma/P_\Gamma^\circ\right)^\gamma \cdots} \tag{3.25}$$

The ratio K_P is extremely useful as it allows the endpoint of a reaction to be determined from the knowledge of ΔG^0.

3.4   Chemical equilibrium in solutions

The preceding section applies to gases that react to produce gaseous products. In practice, however, and especially in foods, the overwhelming majority of reactions take place in solutions. It is preferable to work with concentrations rather than pressures. The reader is reminded that in ideal gases, $PV = nRT \Rightarrow C = n/V = PV/R$, that is, the concentration is proportional to the (partial) pressure.

For solutions, the standard pressures (usually 1 atm) are replaced by standard concentrations: 1 mol dm^{-3} under a pressure of 1 atm. For solids, the standard concentration is defined as the pure solid under a pressure of 1 atm.

For the reaction

$$\alpha A + \beta B + \gamma C \ldots \to xX + yY + zZ + \ldots$$

the equilibrium constant K (without the subscript P, as pressures are not involved) is given by

$$K = \frac{\left([X]/[X]^\circ\right)^x \left([Y]/[Y]^\circ\right)^y \left([Z]/[Z]^\circ\right)^z \cdots}{\left([A]/[A]^\circ\right)^\alpha \left([B]/[B]^\circ\right)^\beta \left([C]/[C]^\circ\right)^\gamma \cdots} \tag{3.26}$$

As $[A]^\circ = [B]^\circ = [C]^\circ = \ldots = 1$ mol dm^{-3}, we can rewrite this as

$$K = \frac{[X]^x[Y]^y[Z]^z \cdots}{[A]^\alpha[B]^\beta[C]^\gamma \cdots} \tag{3.27}$$

where the concentration of A is presented as [A], of B as [B], and so on.

The reaction constant is significantly influenced by the temperature. A very common problem in food processing is the determination of the equilibrium point of a reaction, as the temperature varies in the case of thermal processing (cooking, pasteurization, chilling, freezing). To correlate K with the temperature, we begin with the relationship

$$-\Delta G^\circ = RT \ln K \Rightarrow \ln K = \frac{-\Delta G^\circ}{RT} \qquad (3.28)$$

Differentiating for the temperature,

$$\frac{d(\ln K)}{dT} = -\frac{d\left(\Delta G^\circ/T\right)}{R\,dT} \qquad (3.29)$$

Here we should remember that

$$\Delta G^\circ = \Delta H^\circ - T\Delta S^\circ \Rightarrow \frac{\Delta G^\circ}{T} = \frac{\Delta H^\circ}{T} - \Delta S^\circ \qquad (3.30)$$

Assuming that ΔH° and ΔS° are constant across the range of temperatures that interests us, we differentiate:

$$\frac{d\left(\Delta G^\circ/T\right)}{dT} = -\frac{\Delta H^\circ}{T^2} \qquad (3.31)$$

Substituting the above into Equation (3.29), we obtain

$$\frac{d(\ln K)}{dT} = \frac{\Delta H^\circ}{RT^2} \qquad (3.32)$$

This equation is known as the *van 't Hoff isochore*. It is extremely important as it allows the calculation of the equilibrium constant of a reaction at different equilibria. Considering a change of temperature T_1 to T_2 we can integrate:

$$\int_{K_1}^{K_2} d(\ln K) = \int_{T_1}^{T_2} \frac{\Delta H^\circ}{RT^2} dT \qquad (3.33)$$

from which derives the more functional form of the van 't Hoff equation:

$$\ln \frac{K_2}{K_1} = -\frac{\Delta H^\circ}{R}\left(\frac{1}{T_2} - \frac{1}{T_1}\right) \qquad (3.34)$$

With the above equation, if we know the standard enthalpy of the chemical reaction, we can find at least the percentage increase or decrease of the constant of a reaction between two temperatures. It must again be stressed that the basic premise is that ΔH^0 and ΔS^0 will remain constant between temperatures T_1 and T_2, a premise that is more likely to be met if the two temperatures are close together.

3.5   Ideal solutions: The chemical potential approach

As a case study, let us examine the properties of ideal solutions again, this time using chemical potential instead of free energies. This is a significant exercise, as chemical potential is preferred by many over free energy, and it is good for us to become familiar with this concept. For the chemical potential μ of a vapor at equilibrium, the following relationship applies:

$$\mu_i(g) = \mu_i^o(g) + RT \ln p_i \tag{3.35}$$

When equilibrium between the solution and its vapor is achieved, the chemical potential μ(g) of the vapor is equal to the chemical potential μ(sol) of the solution, so

$$\mu_i(sol) = \mu_i^o(g) + RT \ln p_i \tag{3.36}$$

and from Raoult's law $p_i = p_i^* X_i$:

$$\mu_i(sol) = \mu_i^o(g) + RT \ln X_i p_i^* = \left[\mu_i^o(g) + RT \ln p_i^* \right] + RT \ln X_i \tag{3.37}$$

The above equation applies for every component of an ideal solution. On the right-hand side of Equation (3.37), the first and second factors are the chemical potential of the vapor of pure component *i*. If we say that the component *i* is the solvent, then the bracketed section is equal to the chemical potential of the pure liquid $\mu_i^*(l)$ because, under equilibrium conditions, $\mu_i^*(l) = \mu_i^*(g)$:

$$\mu_i(sol) = \mu_i^*(l) + RT \ln X_i \tag{3.38}$$

Because the solution is obviously not an ideal gas, the above equation does not apply at pressures above the standard reference pressure (1 atm). Despite this, as most reactions that involve food materials take place in conditions close to 1 atm, the chemical potential of a liquid at arbitrary pressure $\mu_i^*(l)$ can be considered approximately equal to the chemical potential at reference pressure $\mu_i^0(l)$.

Consequently, we can write as an approximation

$$\mu_i(sol) = \mu_i^0(l) + RT \ln X_i \tag{3.39}$$

The solutions that follow the above equation are the ideal solutions, defined in Equation (3.11) in terms of free energy. Ideal solutions are the theoretical equivalent to ideal gases for liquid systems: They are encountered only at very low concentrations of dissolved substances when the mutual interactions of the dissolved molecules are practically nonexistent.

3.6   Depression of the freezing point and elevation of the boiling point

How is the freezing of a liquid influenced by the presence of another substance dissolved in it? Consider a pure liquid at a temperature equal to its freezing point T_{fr}. The material freezes because the free energies of the liquid and solid phases are equal ($G_{liq} - G_{solid} = 0 \therefore \Delta G = 0$). The addition of a substance to this liquid will lower the free energy of the system by $\Delta G = RT \ln X_1$, where X_1 is the mole fraction of the solute in the solution. This means that the system will no longer be in equilibrium, which will occur at a new temperature T'_{fr}, where $\Delta G = 0$. What is the new temperature at which the solution freezes? Let us remember the equation

$$\frac{d(\Delta G/T)}{dT} = -\frac{\Delta H}{T^2} \tag{3.40}$$

and let us integrate the second part from the freezing point T_{fr} of the solution (where $\Delta G = RT\ln X_1$) to the new freezing point T'_{fr} that results from the dissolution of the substance (where $\Delta G = 0$), from which we get

$$\int_{\Delta G = RT \ln X_1}^{\Delta G = 0} d\left(\frac{\Delta G}{T}\right) = -\int_{T_{fr}}^{T'_{fr}} \frac{\Delta H}{T^2} dT \Rightarrow$$

$$R \ln X_1 = -\Delta H \left(\frac{1}{T'_{fr}} - \frac{1}{T_{fr}}\right) \Rightarrow \ln X_1 = -\frac{\Delta H}{R}\left(\frac{T_{fr} - T'_{fr}}{T_{fr}T'_{fr}}\right) \tag{3.41}$$

If the depression of the freezing point is very small (as is generally the case), we can accept that $T'_{fr} \approx T_{fr}$, so $T'_{fr}T_{fr} \approx T_{fr}^2$. For systems with two components (X_1 and X_2), we can say that $1 = X_1 + X_2 \therefore X_1 = 1 - X_2$:

$$\Delta T_{fr} = -\frac{\ln(1 - X_2) RT_{fr}^2}{\Delta H_{fus}} \tag{3.42}$$

Furthermore, if $X_2 \ll 1$ (e.g., small quantities of salt in a large volume of water), $\ln(1 - X_2) \approx -X_2$.

On this basis, Equation (3.16) can be written as

$$\Delta T_{fr} = \frac{X_2 R T_{fr}^2}{\Delta H_{fus}} \tag{3.43}$$

The enthalpy ΔH_{fus} is called the *latent heat of fusion*.

Similarly for the elevation of the boiling point resulting from the addition of a substance to a solvent (in a low concentration), we have

$$\Delta T_{vap} = \frac{X_2 R T_{vap}^2}{\Delta H_{vap}} \tag{3.44}$$

where T_{vap} is the boiling point of the pure solvent and ΔH_{vap} is the latent heat of vaporization.

With the above equations it is possible to calculate the depression of the freezing point ΔT_{fr} or the elevation of the boiling point ΔT_{vap} of dilute solutions using the appropriate latent heat.

3.7 Osmotic pressure

From the destabilization of dairy products by polysaccharides to the regulation of cellular permeability, and from the production of compost to the transfer of salt from the brine to the cheese, the results of the phenomenon of osmotic pressure are fundamental in food chemistry and food technology.

What is osmotic pressure? To start with, it is pressure, that is, a fundamental property defining the state of material bodies. This pressure is applied to the molecules of a solvent because of the existence of molecules of a solute inside the solvent. Let us think of a vessel that is divided in two by an impermeable barrier. On the right-hand side is a solution of a substance, while on the left is the pure solvent. If the dividing barrier is removed, the solute molecules will move into the pure solvent in order to increase the entropy of the system (it has already been discussed that the greatest entropy is achieved when the molecules of a dissolved substance are evenly distributed throughout the whole of the space available to them). Let us repeat the above experiment, this time with a semipermeable membrane that separates the two halves of the vessel. This membrane is permeable to the solvent but not to the molecules of the dissolved substance (Figure 3.4). Because the solute molecules cannot pass the membrane in order to maximize the entropy, the system reacts by doing the reverse: Molecules of solvent move from the pure solvent into the solution,

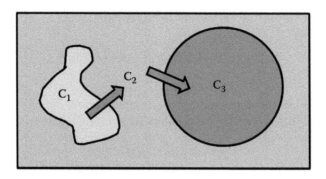

Figure 3.4 Behavior of salt solutions of different concentrations C_x ($C_3 > C_2 > C_1$) separated by semipermeable membranes. Note how the liquid flows (as shown by the arrows) in order to equalize the concentrations.

in order to dilute the dissolved substance. This arrangement of the two liquids with a semipermeable membrane can be likened to a pump that drives the solvent through the membrane. The counter-pressure that must be exerted on the solution in order to stop the operation of this pump is called *osmotic pressure*.

If there is pure solvent or if there are equal concentrations of the solute in both sections of the vessel, then the system is in equilibrium ($\Delta G_{\text{soln}} = 0$). If a solute is added at a mole fraction of X_2 in one of the two compartments (with the mole fraction of the solvent being X_1), the free energy of that compartment will increase by $RT \ln X_1$, while in the other it will reduce by the same amount. For the compartment with the solution, we can write

$$G_{\text{soln}} = G^{\circ}_{\text{soln}} + RT \ln X_1 \tag{3.45}$$

Let us say that we apply a pressure π that is just sufficient to stop the flow of the solvent toward the solution. At equilibrium, the balance of energy must be zero and equals the difference [Change in free energy due to the application of osmotic pressure to a value of G'_{soln}] – [Change in free energy according to $RT \ln X_1$ due to the dilution of the solution]. The mathematical expression of this, integrated from 0 to π, is

$$\int_0^{\pi} dG'_{\text{soln}} + RT \ln X_1 = 0 \tag{3.46}$$

We saw in previous chapters that the change in free energy under constant temperature ($dT = 0$) is given by

$$dG = VdP - SdT = VdP \tag{3.47}$$

If the molar volume of the solvent is V_1, we have

$$\int_0^\pi V_1 dP = -RT \ln X_1 \Rightarrow V_1\pi = -RT \ln X_1 \tag{3.48}$$

If, as we saw in the depression of the freezing point, for $X_2 \ll 1$, one can assume that $\ln(1-X_2) \approx -X_2$. We can thus write for $\ln(1-X_2) \approx -X_2$ and $V_1 = V$:

$$V_1\pi = -RT \ln(1-X_2) = RTX_2 \tag{3.49}$$

Solving for π results in

$$\pi = RT\frac{X_2}{V_1} = RTm \tag{3.50}$$

where m is the molarity ([mol substance] [dm^3 solution]$^{-1}$) of the dissolved substance.

3.8   Polarity and dipole moment

Later in the text there is extensive discussion of polar and nonpolar molecules; polarity has already been alluded to in the discussion of hydrophobic interactions between fatty acids in Section 2.4.1. The rules of practical chemistry dictate that substances of a particular polarity will dissolve more readily in a solvent of comparable polarity. The polarity of a molecule is quantified using the molar polarity P, which is given by the Debye equation:

$$P = \frac{\varepsilon-1}{\varepsilon+2}\frac{M}{d} = \frac{4}{3}\pi N\left(a + \frac{\mu^2}{3kT}\right) \tag{3.51}$$

where d is the density of the material that is composed of molecules of molecular mass M.

The factor ε is the dielectric constant of the material and is equal to the ratio of the capacity of a capacitor with the liquid in question as a dielectric (insulator) to its capacity with air as a dielectric. The last part of the equation includes the molecular dipolar torque μ (a direct measure of polarity), the polarizability of the molecule a (the dipolar moment that develops on exposure to an electrical field), Avogadro's number N, and the Boltzmann constant k.

The dipole moment is equal to the vector sum of the electrical charge times the distance of the charged poles of the molecule. If the charge is symmetrically distributed around the center of the molecule (e.g., in

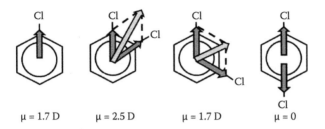

$\mu = 1.7\,D$ $\qquad$ $\mu = 2.5\,D$ $\qquad$ $\mu = 1.7\,D$ $\qquad$ $\mu = 0$

Figure 3.5 Polarity of a series of chlorine-substituted benzenes (chlorobenzene; *ortho-*, *meta-*, and *para-*dichlorobenzene). The dipolar moment of each compound is written below. Note how the vector of the dipolar moment is directed from the electropositive center to the electronegative extremity.

methane, *para*-dichlorobenzene (Figure 3.5), tetrachloromethane, or a very large polymethyl chain), the vector sum of the charges is close to zero. These molecules have a very low dipole moment and are called *nonpolar*. If, on the other hand, the vector sum of the charges is a large number (e.g., in nonsymmetrical ions, acetonitrile, or water, which is not a symmetrical molecule because the H–O–H angle is approximately 105°), then the molecules are termed *polar*. The significance of this for the forthcoming discussion lies in the fact that electronegative substituents (e.g., halogens, oxygen, nitrogen, and charged groups in general), when they are not arranged with total symmetry on an organic molecule, and especially when they are bound with hydrogen and/or carbon, greatly increase the polarity of the molecule in question. On the other hand, extended aliphatic or aromatic carbon chains generally tend to have a low dipole moment. It is important to remember that polar and nonpolar substances cannot be mixed easily.

3.8.1 Polarity and structure: Application to proteins

Proteins are macromolecules with molecular weights that are usually on the order of kilo-Daltons (kDa) or even mega-Daltons (Mda). From the chemical point of view, proteins are *polypeptides*, that is, polymers of α-amino acids (amino acid – amino acid – amino acid – amino acid –... –amino acid). The amino acids are compounds that contain an amino group (–NH$_2$, –NH–) and a carboxyl group (–COOH). More particularly, α-amino acids are those in which the amine group and the carboxyl group are substituents on the same carbon atom. The α-carbon atom (that to which the amino group and the carboxyl group are joined) of an amino acid has at least a hydrogen atom as the third substituent. The amino acids are divided into basic, neutral, and acidic on the basis of their substituent side-chain. In basic amino acids, the substituent contains a weak base (i.e., a second amino group), while the substituent of

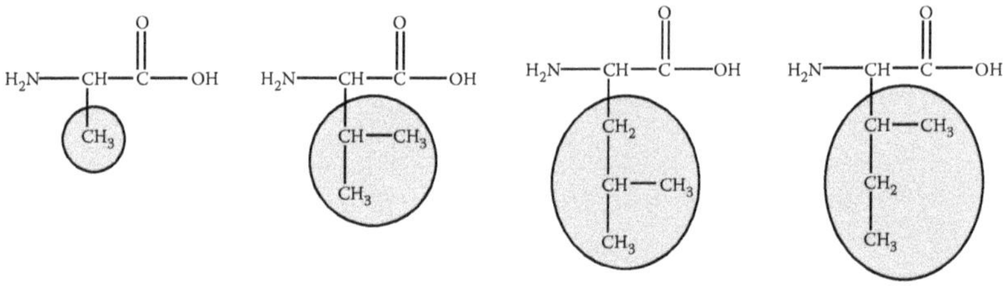

Alanine   Valine   Leucine   Isoleucine

Figure 3.6 Four amino acids sorted with increasingly hydrophobic character. Notice the relation between ranking and the highlighted parts.

acidic amino acids contains a weak acid (i.e., a second carboxyl group, or a simple hydroxyl group in the case of tyrosine, or a sulfhydryl group in the case of cysteine). The physicochemical behavior of amino acids is directly related to the substituent group on the α-carbon. Let us consider the amino acids in terms of their solubility in water as compared to their solubility in less polar solvents, that is, their free energy of transfer from water to ethanol $\Delta G_{trans(wat\text{-}eth)}$, which relates to the reaction (amino acid dispersed in ethanol → amino acid dissolved in water), or other similar parameters (free energy of transfer from nonpolar to more polar solvents, or vice versa).

Although the details of various experiments may differ, a common ground is that the less polar amino acids will transfer more readily from water to less polar environments. An example is depicted in Figure 3.6 in the form of the series alanine – valine – leucine – isoleucine. An increase in hydrophobicity from the addition of a methyl group makes the transfer of the amino acid to a more polar phase more difficult. In contrast, for more polar groups (e.g., serine, glutamine), transfer from a less polar solvent to a more polar one is favored. It would be expected therefore that the polar amino acids will arrange themselves on the outside of a protein chain with the less polar amino acids located internally (away from the water).

3.9   *Real solutions: Activity and ionic strength*

Raoult's equation (Equation (3.1)) shows the dependence of the vapor pressure p^* of a substance on its molar fraction X in a solution of combined total vapor pressure p. In reality, the mutual interactions between the molecules of a solution are not homogeneously distributed, and the forces between random pairs of molecules in the solution are not always the same. We have already seen deviations from Raoult's law in the case of azeotropic mixtures, which are due to the development of forces between the individual components of a mixture.

In practice, the vapor pressure of a substance is a function not so much of its molar fraction (as predicted by Equation (3.1)) but of a quantity called *activity a*:

$$p = p^* a \qquad (3.52)$$

where *a* is the product of the mole fraction and a corrective factor γ that is called the *activity coefficient*:

$$a = \gamma X \qquad (3.53)$$

The divergence of *a* and *X* is particularly significant in solutions of electrolytes (substances that ionize). This is because the charges on the molecules interact, exerting additional forces on the molecules of the system.

The total effect of charge on the behavior of an aqueous system is given by the ionic strength *I* of the solution, which is equal to

$$I = 0.5 \ \Sigma(c_i Z_i^2) \qquad (3.54)$$

where c_i is the concentration of substance *i* and Z_i its charge.

With the assumption that any deviations from ideal behavior of electrolyte solutions are due to electrostatic effects, Debye and Hückel derived the equation below for the calculation of the total activity coefficient $\gamma_\pm$ for a solution with ionic strength *I* containing cations of charge Z_+ and anions of charge Z_-:

$$\log \gamma_\pm = -0.5 Z_+ Z_- \sqrt{I} \qquad (3.55)$$

The Debye and Hückel equation refers to solutions at a temperature of 298K. At different temperatures, the coefficient 0.5 in Equation (3.28) takes different values. Significant deviations can occur, however, between the calculated and the real values.

The concept of ionic strength is useful because it quantifies the effect of aqueous solution on the behavior of dissolved salts, acids, and bases. By the same reasoning, pH can be defined as the reverse decimal logarithm of the activity and not of the concentration of hydrogen ions in aqueous solution. Generally speaking, it is useful for activity to be used rather than concentration in electrolyte solutions (in general, most solutions that concern foods). This can be difficult however, as the values of γ and *I* are not easy to calculate.

3.10   On pH: Acids, bases, and buffer solutions

Acidity and alkalinity are two fundamental parameters in most issues pertaining to food and to biological systems in general. These are connected to the presence and concentration (or, better yet, the activity) of hydrogen H^+ or hydronium cations H_3O^+, and of hydroxyl anions OH^-. Hydronium is the result of the reaction of a hydrogen cation with a molecule of water; in reality, complexes of protons with more than one water molecule are also possible; here, hydronium is used for simplicity. The dissociation of water can be written as

$$H_2O \rightleftharpoons H^+ + OH^-$$

$$2H_2O \rightleftharpoons H_3O^+ + OH^-$$

Protonation of water $H_2O + H^+ \rightarrow H_3O^+$ is generally treated as one-way, and it is not unusual for hydronium ions to be used instead of hydrogen cations in water dissociation calculations. Assuming that the solutions are dilute enough for the activity to be equal to the concentration, the dissociation constant is $K = [H^+][OH^-]/[H_2O]$. Considering that the concentration of nondissociated water $[H_2O]$ is far greater than the concentration of the ions resulting from this dissociation (and hence its concentration can be considered unchanged), this constant can be written as $K = [H^+][OH^-]$; that is, the product of the two concentrations remains unchanged. At 298 K, this product $[H^+][OH^-]$ will be 10^{-14}: when the concentration of [H+] increases, the concentration $[OH^-]$ will decrease so as to satisfy the condition $[H^+][OH^-] = 10^{-14}$. This means that one species of either hydrogen/hydronium ions or hydroxyls will predominate. The single case where hydrogen/hydronium ions exist at the same concentration as the hydroxyls will be at 10^{-7} mol dm^{-3} for each ($10^{-7} \times 10^{-7} = 10^{-14}$). The concept of pH was developed in order to express the concentration of hydrogen ions (and hence indirectly that of the hydroxyls) without the need to revert to large, negative exponents. The pH value of a solution is defined as

$$pH = -\log[H^+] \tag{3.56}$$

In the above-mentioned scenario, $[H^+] = [OH^-] = 10^{-7}$ mol dm^{-3}, the pH should be equal to $-\log_{10}[10^{-7}] = 7$. Such solutions are called *neutral*. In the case of a predominance of hydrogen cations, for example, when $[H^+] = 10^{-3}$ mol dm^{-3} (a ten thousand times higher concentration of hydrogen ions compared to the previous case), $-\log_{10}[10^{-3}] = 3$. Such solutions are called *acidic*, as it is the dissociation of acids that produces significant concentrations of hydrogen/hydronium cations. In this case, the concentration of hydroxyls will be $10^{-3}[OH^-] = 10^{-14}$ thus $[OH^-] = 10^{-11}$ mol dm^{-3}. In another scenario, let us assume that, due to the addition of a base, hydroxyls have increased in concentration to $[OH^-] = 10^{-4}$ mol dm^{-3}, bringing the concentration of hydrogen cations to $[H^+]10^{-4} = 10^{-14}$ thus $[H^+] = 10^{-10}$. pH will then be $-\log_{10}[10^{-10}] = 10$. These solutions are called *basic* or *alkaline*, as they result from the dissociation of alkaline substances and of bases in general. Here we should remember that the above are valid in dilute solutions, where the activity can be considered practically equal to the concentration. Additionally, because all of the above are built on the hypothesis of dissociating water, and negating the $[H_2O]$ term, they only apply to purely aqueous systems.

It is easy to understand that pH 7, corresponding to $[H^+] = [OH^-] = 10^{-7}$ mol dm^{-3}, is the midpoint on a scale in which pH values higher than 7 correspond to basic environments, while pH values lower than 7 correspond to acidic ones. Here, we should briefly discuss acids and bases. In a simple, intuitive definition, one would consider an acid to be a substance that releases H^+ (essentially protons), while a base would be a substance that releases OH^-. Brønsted and Lowry expanded this by defining an acid as a substance that releases protons (as previously) and a base as a substance that *accepts* protons. That means that an acid, after its dissociation, becomes a base (called the acid's *conjugate* base) because it is now lacking a proton, and vice versa.

A substance that dissolves into water and releases hydrogen cations ("protons") is by all definitions an acid. Let us consider its dissociation:

$$HA \rightleftharpoons A^- + H^+$$

Or, considering that the hydrogen will end up as a hydronium ion:

$$HA + H_2O \rightleftharpoons A^- + H_3O^+$$

Complete dissociation of 1 mol of any such acid would produce 1 mol of hydronium ions. However, not all HA will be dissociated. Assuming dilute aqueous systems, the constant of the dissociation reaction can be written as

$$K_a = \frac{[A^-][H_3O^+]}{[HA]} \tag{3.57}$$

$[H_2O]$ does not partake in the equation, as we can assume that its concentration is too large and does not change during the reaction. Large values of K_a stand for high values of $[H_3O^+]$—that is, for strong acids. Similar to pH, one can define the negative decimal logarithm of K_a as $pK_a = -\log_{10} K_a$. In this case, the smaller the value of pK_a, the stronger the acid. By the above, a strong acid is one that can produce more hydronium ions for the same initial concentration of acid HA, that is, bring the pH to a lower value.

Another property of strong acids is that they show substantial dissociation even at low pH values: A close examination of the equation

$HA + H_2O \rightleftharpoons A^- + H_3O^+$ suggests that, according to Le Chatelier's principle, an increase in the concentration of $[H_3O^+]$ (i.e., lowering of the pH) favors the nondissociated form, that is, shifts this reaction toward the reagents. In this context, the strongest of two acids is the one that, for the same initial concentration $[HA]$, dissociates to a larger degree at the same overall $[H_3O^+]$ (or pH). According to Equation (3.57), a larger dissociation for the same value of $[H_3O^+]$ implies a larger dissociation constant K_a and a relatively low pK_a. The latter can thus be used in order to assess the acid's relative strength, lower pK_a standing for stronger acids. In a similar manner, K_b and pK_b can be defined for the dissociation of bases.

The effect of the degree of dissociation of an acid on the solutions pH can be calculated, to a first approximation, by solving Equation (3.57) for the hydronium ion (or hydrogen cation) concentration and taking the logarithms:

$$K_a = \frac{[A^-][H_3O^+]}{[HA]} \Rightarrow [H_3O^+] = K_a \frac{[HA]}{[A^-]} \Rightarrow pH = pK_a - \log_{10} \frac{[HA]}{[A^-]} \quad (3.58)$$

The final expression above is called the *Henderson–Hasselbalch equation*. Assuming that a base is added into a solution of weak acid, neutralization will result in changes in pH. The pH will change with base addition and the subsequent neutralization. However, according to the Henderson–Hasselbalch equation, when $[HA] = [A^-]$, the logarithm of their ratio is zero, and thus their contribution to Equation (3.58) is minimal. In that case, $pH = pK_a$. That suggests that the coexistence of a weak base and its conjugate acid, or vice versa, will make the systems somehow resist changes in pH. These systems are called *buffer solutions*, and they play a major role in controlling the pH of biological systems. As suggested by the Henderson–Hasselbalch equation, the maximum buffering capacity of a weak acid and its conjugate base is achieved when the pH is equal to the pK_a of the weak acid. The above also applies to weak bases and their conjugate acids.

What happens in the case of a *polyprotic* acid, that is, an acid able to dissociate two or more times, releasing two or more hydrogen cations? All these protons are not released simultaneously. An equilibrium is established in the form

$$H_nA \rightleftharpoons H_{n-1}A^- + H^+ \rightleftharpoons \dots \rightleftharpoons A^{-n} + H^+$$

Each of the consecutive dissociations has its own reaction constant K_{a1}, K_{a2}, etc.:

$$K_{a1} = \frac{[H_{n-1}A^-][H^+]}{[H_nA]} \quad (3.59)$$

$$K_{a2} = \frac{\left[H_{n-2}A^{2-}\right]\left[H^+\right]}{\left[H_{n-1}A^-\right]}$$

(3.60)

and its equivalent pK_{ai} value. Because the dissociation of H_nA is required before dissociation of $H_{(n-1)}A^-$ takes place and so on, pK_{an} has the highest value and pK_{a1} the lowest. The above are of fundamental importance to the charge and conformation of proteins, but also of charged polysaccharides and a large array of other macromolecules.

In the case of coexistence of basic and acidic groups attached to the polymer chain (the acidic and basic amino-acids comprising a protein are typical examples of this scenario), the acidic groups will tend to dissociate and become negatively charged at high pH (i.e., in an alkaline environment), while the bases will dissociate and become positively charged at low pH (i.e., in an acidic environment). This suggests that, during the titration of a protein from high to low pH and vice versa, a pH value will be attained where the two charges are equal, that is, the entire protein has a net zero charge. This pH value is called the protein's *isoelectric point* pI.

3.11   *Macromolecules in solution*

So far, the discussion has evolved around molecules in solution. An unwritten assumption has been that the dissolved molecules were *small*, or to put it better, comparable in size with the molecules of the solvent. In a simple approach, the molecules of a solution comprised of substance A dissolved in substance B can be visualized as circles embedded in the grid of Figure 3.7, left-hand side. Each slot in this grid is occupied either by a dark-colored solute molecule or an open-colored solvent molecule. Let us assume that any physical interactions between a solvent and a solute molecule are roughly equal to the interactions between two solvent molecules or between two solute molecules. Individual molecules of the dissolved substance can move, for example due to thermal motion or due to a shear field. Such a movement would cause them to occupy a slot previously occupied by another molecule (Figure 3.7, right-hand side). In the case of thermal motion, this exchange of positions would increase the system's entropy S, as it would favor the maximal dispersion of molecules A into B.

Another assumption incorporated in the previous rationale is that, although these molecules could be asymmetrical, they must be

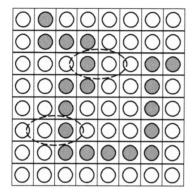 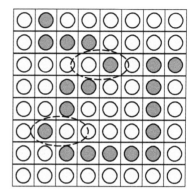

Figure 3.7 Exchange of grid cells for solvent and solute, leading to an increase in entropy.

rigid nevertheless; at least they should be considered rigid for a simple description of a system: That is, they do not alter their shape significantly during their residence in the solvent matrix. However, these assumptions are only partially correct in the components of real food systems. For example, even simple molecules change their shape by means of rotation of their single covalent bonds. This has significant consequences in the shape of molecules. Things become even more complicated when the size of the dissolved molecule becomes hundreds or thousands of times larger than that of the solvent. Polysaccharides, for example, are entities made of tens or hundreds of simple sugars joined together by means of covalent bonds. Such molecules, often called *macromolecules*, are basically simple molecules making up larger entities by means of forming covalent bonds between them.

3.12   *Enter a polymer*

The simple model described previously involves individual, free-moving molecules. Their random thermal motion (which gives rise to entropy) leads them to move in different directions, irrespective of their neighbor's motion. What would change if we permanently linked adjacent molecules of substance A (all dark circles) via strong and permanent covalent bonds? In that case, as depicted in Figure 3.8, the momentum of a moving particle would be transferred to its neighbors *via* the covalent bonds; the same would apply for all components of the newly defined *macromolecular chain*. Not only can the individual molecules not move independently, but in the case that they move in opposite directions, their momentum is mutually eliminated; if they move in different directions, direct transfer of momentum between the two components can lead to circular-like movement.

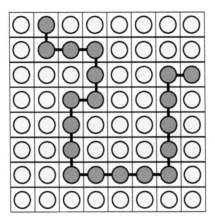

Figure 3.8 Visualization of a polymer embedded in a two-dimensional grid, surrounded by solvent molecules.

Zooming out, entire parts of the chain have their movement negated or altered due to their interaction with their similar neighboring parts. This system is, in practice, a very large molecule, appropriately called a *macromolecule* or a *polymer* from the Greek words *polý* ("a lot") + *méros* ("part"), in a solvent comprised of smaller molecules (light gray circles in Figure 3.7). This is a far more complex physicochemical entity.

3.13   Is it necessary to study macromolecules in food and biological systems in general?

Yes it is, as some of the most important components of food systems, such as proteins and polysaccharides, are macromolecules. The same applies for nucleic acids. Even triglycerides, the bulk of most oils and fats, could be considered polymer melts or polymer crystals accordingly when they are comprised of large fatty acids. As we have seen in previous paragraphs, in ideal solutions the molar fraction is equal to the activity; that activity can be used in Raoult's law instead of the molar fraction. As the molecules of solute become far larger than those of the solvent, Raoult's law and the ideality of simple solutions cease to apply. In subsequent paragraphs we try and build a thermodynamic image of a polymer in solution and use that to investigate the conditions under which a polymer dissolves or phase separates in a solvent. This is not a theoretical conversation, but an essential step in understanding why molecules of such tremendous complexity as proteins, polysaccharides, and glycoproteins exist in foods in the first place, what their structural and functional roles are, and what is their particular contribution in food processing in eating and in digestion.

3.13.1   Intrinsic viscosity

The specific viscosity η_{sp} of a polymer solution η_{pol} can be defined as the fractional change in viscosity upon addition of an amount of polymer:

$$\eta_{sp} = \frac{\eta_{pol} - \eta_{solvent}}{\eta_{solvent}} \tag{3.61}$$

where $\eta_{solvent}$ is the initial viscosity of the solvent. As the viscosity of a polymer solution depends on the amount and conformation of the polymer, information on the intrinsic properties of the polymer can be estimated by means of extrapolation of the solvent's viscosity to a polymer concentration c of zero; that is, take the η_{pol} limit for $c \to 0$. This is called the polymer's *intrinsic viscosity* $[\eta]$.

$$[\eta] = \lim_{c \to 0} \frac{\eta_{sp}}{c} = \lim_{c \to 0} \frac{\eta_{pol} - \eta_{solvent}}{c\eta_{solvent}} \tag{3.62}$$

The units of the intrinsic viscosity are volume over mass; one of its usual expressions is dL g^{-1}. It is important to notice here that, despite its name, intrinsic viscosity is *not* viscosity but volume per unit mass. It is convenient to consider that a dissolved polymer's intrinsic viscosity expressed in dL g^{-1} is the volume in dL occupied in solution by 1 g of polymer material. In that respect, its inverse $[\eta]^{-1}$, now in units of mass per unit volume, should be somehow related to a system where all the solvent is occupied by uncompressed polymer molecules. This description stands close to that to be soon discussed for the semidilute regime. The relationship between a polymer's molecular weight M and its intrinsic viscosity is described by the *Mark–Houwink equation*:

$$[\eta] = KM^a \tag{3.63}$$

The *Mark–Houwink* equation is useful, especially in terms of the exponent a. Its determination is usually attained by means of constructing a $\log[\eta] - \log M$ plot and determining its slope a. Small values of a, that is, ~0.5, suggest a flexible polymer in a close-to-ideal solvent; values between 0.5 and 0.8 suggest a good solvent; while higher values suggest a stiffer chain and decreasing solvent quality.

3.14   Flory–Huggins theory of polymer solutions

An elegant approach to describe the dissolution of polymers was proposed by Flory and by Huggins, and today it forms the basis of our understanding of the subject. Flory and Huggins described the process of dissolution by considering that the polymer is initially in a "solid" form. This can be visualized as a rigid conformation of a polymer being part of a perfectly oriented ensemble of macromolecules, directly adjacent to each other. This hampers the mobility of the polymer chain, sterically inhibiting any changes in its conformation. Dissolution is described in two steps: first, as a removal of the polymer from its immediate neighborhood of the other polymers (thus allowing it to increase its possible conformations and hence its degrees of freedom), and then the solvation of its component monomers by molecules of the solvent. The first step is an "entropic process," as it pertains to the increase in the degrees of freedom in monomer movement, while the second is also an "enthalpic" process, as it involves establishment of direct interactions between solvent and solute.

The calculation of the free energy of mixing ΔG^M, which will eventually tell us if the polymer dissolution is spontaneous (i.e., thermodynamically favored), involves the separate calculation of the entropy of mixing ΔS^M and the enthalpy of mixing ΔH^M, which are discussed in Sections 3.14.1, 3.14.2, and 3.14.3.

3.14.1 Conformational entropy and entropy of mixing

Let us assume a polymer population consisting of similar polymers, each one comprised of r monomers that form a continuous chain via covalent bonds between them. As a first approximation, we can take the size of the monomer to be similar to that of the solvent. Let us also assume a three-dimensional grid, similar to the two-dimensional one depicted in Figures 3.7 and 3.8. Only one molecule, either a monomer or a solvent molecule, can fit into each cell.

Let us now concentrate on the first question: In how many distinct ways can we place a macromolecular chain into this grid? A first limiting factor is the need for the monomer-occupied cells to be adjacent, as the monomers contained therein would be connected with covalent bonds in order to form the macromolecular chain. Assuming a total cell count N_o, each containing either a solvent or a monomer molecule (all the cells are occupied; no void exists), let us examine the mixing of N_1 solvent molecules with N_2 polymers. Assuming a monodisperse polymer population (all polymers are of the same size), the number of monomer segments is equal to rN_2.

$$N_o = N_1 + rN_2 \tag{3.64}$$

Incorporation of the macromolecule into the grid involves the placement of the polymers one by one. After i macromolecules have been incorporated into the matrix, the number of cells still available for the placement of macromolecule $i + 1$ is equal to $N_o - ri$. Insertion of macromolecule $i + 1$

can begin by placing one of the two terminal monomers into a vacant cell. The monomer connected to this must be placed into an adjacent empty cell. Let us define as *coordination number z* the number of direct neighboring cells of each cell in the grid. For example, the coordination number for each cell in the two-dimensional grid presented in Figure 3.8 is 4, while, of course, this can be far larger in a three-dimensional, noncubic grid. Let us consider a possibility p_i for an adjacent cell to be vacant. Assuming that z is sufficiently large, one may consider that p_i equates to the fraction of the void cells of the entire matrix:

$$p_i = \frac{N_o - ri}{N_o} \tag{3.65}$$

thus bringing the number of immediately neighboring vacant cells to zp_i. Because one cell is always occupied by the previously inserted monomer molecule, the number of available cells for every subsequent cell is $(z - 1)p_i$. This brings the number of ways $W_{(i+1)}$ under which polymer $i + 1$ can fit into the matrix down to

$$W_{(i+1)} = (N_o - ri)z(z-1)^{r-2}\left(\frac{N_o - ri}{N_o}\right)^{r-1} = (N_o - ri)^r\left(\frac{z-1}{N_o}\right)^{r-1} \tag{3.66}$$

while the number of the ways that all polymers can be incorporated into the matrix W_p is equal to

$$W_p = \frac{W_1 W_2 \ldots W_i \ldots W_{N_2}}{N_2!} = \frac{\prod_{i=1}^{N_2} W_i}{N_2!} \tag{3.67}$$

Combining the previous two equations yields

$$\tag{3.68}$$

$$W_p = \frac{\prod_{i=1}^{N_2} W_i}{N_2!} = \left(\frac{z-1}{N_0}\right)^{(r-1)N_2} \frac{\prod_{i=1}^{N_2}\left[N_0 - r(i-1)\right]^r}{N_2!}$$

Multiplying by and dividing by r yields

$$\prod_{i=1}^{N_2}\left[N_o - r(i-1)\right]^r = r^{(N_2 r)}\prod_{i=1}^{N_2}\left(\frac{N_o}{r} - i + 1\right)^r \qquad (3.69)$$

Taking into account that expression

$$\left(\frac{N_o}{r}+1-1\right)^r\left(\frac{N_o}{r}+1-2\right)^r\left(\frac{N_o}{r}+1-3\right)^r\cdots\left(\frac{N_o}{r}+1-N_2\right)^r \qquad (3.70)$$

can be equated to

$$\left[\frac{\left(\frac{N_o}{r}\right)!}{\left(\frac{N_1}{r}\right)!}\right]^r \equiv F \qquad (3.71)$$

Equation (3.60) can be re-written as

$$\qquad\qquad (3.72)$$

$$W_p = \frac{Fr^{rN_2}}{N_2!}\left(\frac{z-1}{N_o}\right)^{(r-1)N_2}$$

The remaining cells will be filled with solvent molecules. Because all solvent molecules are similar and indistinguishable, substitution of the void in all empty cells for solvent molecules does not contribute to the overall entropy. And because the overall entropy depends only on the placement and conformation of the macromolecules, the possibility for a situation P to occur in the overall matrix is equal to the number of ways that all polymers can be incorporated into the matrix W_p.

$$S^M = k\ln P \equiv S^M = k\ln W_p \qquad (3.73)$$

Here we should briefly introduce the Stirling approximation, according to which the possible combinations of w nonrepeated discrete objects per r objects are equal to

$$w! = \left(\frac{w}{e}\right)^w\sqrt{2\pi w} = w^{w+\frac{1}{2}}e^{-w}\sqrt{2\pi} \qquad (3.74)$$

which, for very large numbers, becomes $\ln w! = w\ln w - w$.

The entropy of the system can thus be expressed as

$$-\frac{S^M}{k} = -\ln W_p = N_1 \ln\left(\frac{N_1}{N_o}\right) + N_2 \ln\left(\frac{N_2}{N_o}\right)$$

$$+ N_2\left[(r-1)\ln(z-1)-(r-1)\right]$$

(3.75)

Adding and subtracting $N_2 \ln r$ on the right-hand side of Equation (3.67), we can obtain the more useful form:

$$\frac{S^M}{k} = -N_1 \ln\left(\frac{N_1}{N_1 + rN_2}\right) - N_2 \ln\left(\frac{rN_2}{N_1 + rN_2}\right) + N_2\left[(r-1)\ln\frac{(z-1)}{e} + \ln r\right]$$

$$= -N_1 \ln\varphi_1 - N_2 \ln\varphi_2 + N_2\left[(r-1)\ln\frac{(z-1)}{e} + \ln r\right] \therefore$$

(3.76)

$$S^M = -k\left\{N_1 \ln\varphi_1 + N_2 \ln\varphi_2 - N_2\left[(r-1)\ln\frac{(z-1)}{e} + \ln r\right]\right\}$$

where

$$\ln\phi_1 = \ln\frac{N_1}{N_1 + rN_2}$$

(3.77)

$$\ln\phi_2 = \ln\frac{N_2}{N_1 + rN_2}$$

(3.78)

This entropic quantity is only related to the random distribution of the polymer molecules into the matrix. The coexistence of two populations of molecules (solvent and solute) gives rise to two more entropic quantities, this related to the dispersion of solvent molecules into solute and vice versa. For pure solvent ($N_2 = 0$), as discussed previously, the entropy S_1 equals zero, as all cells are occupied by identical molecules. For pure polymer ($N_1 = 0$), the entropy S_2 is a non-zero number and is related to the entropy of the randomly placed macromolecules. It can be calculated from Equation (3.76):

$$S_2 = kN_2\left[(r-1)\ln\left(\frac{z-1}{e}\right) + \ln r\right]$$

(3.79)

The entropy of mixing for these randomly arranged molecules is equal to

$$\Delta S^M = -k\,(N_1\,\ln\varphi_1 + N_2\,\ln\varphi_2) \tag{3.80}$$

So far, the discussion has revolved around macromolecules, while the equations involved have mostly been monomers. In order to express our equations in terms of molar quantities of polymers, let us consider our system comprised of n_1 solvent molecules and n_2 polymer (not monomer!) molecules.

Considering that out of a total volume V, V_1 is the volume fraction of solvent and V_2 the volume fraction occupied by the polymer, the following apply for n_1 and n_2 molecules of solvent and polymer, respectively, assuming that during the mixing of solvent and polymer the total volume does not change:

$$\varphi_1 = \frac{n_1 V_1}{n_1 V_1 + n_2 V_2}\,;\ \ \varphi_2 = \frac{n_2 V_2}{n_1 V_1 + n_2 V_2} \tag{3.81}$$

As the discrepancy in sizes between solvent molecules and polymer macromolecules is very large, it is convenient to define the molar volume they both occupy in terms of an abstract reference volume V_{ref}. In this case, the relationships between the molar volume of the solvent and that of the polymer now become $V_1 = r_1\,V_{ref}$ and $V_2 = r_2\,V_{ref}$, respectively. Assuming that the volume does not change during mixing of the components, the total volume is defined as $V = V_1 + V_2 = (n_1 r_1 + n_2 r_2)\,V_o$. The total number of moles in the sample is n and $r_1 + r_2 = r$. Substituting Equation (3.81) into Equation (3.80), we get

$$
\begin{aligned}
\frac{\Delta S^M}{R} &= -V\left(\frac{\varphi_1}{V_1}\ln\varphi_1 + \frac{\varphi_2}{V_2}\ln\varphi_2\right)\\[2mm]
&= -\frac{V}{V_o}\left(\frac{\varphi_1}{r_1}\ln\varphi_1 + \frac{\varphi_2}{r_2}\ln\varphi_2\right)\\[2mm]
&= -rn\left(\frac{\varphi_1}{r_1}\ln\varphi_1 + \frac{\varphi_2}{r_2}\ln\varphi_2\right)
\end{aligned}\tag{3.82}
$$

For small molecules such as the solvent, r_1 tends to be close to unity. For polymers, however, r tends to be a very large number; thus for the last two terms of Equation (3.82), the first tends to be much larger than the second. The importance of this observation is that the entropy of mixing of a polymer into solvent is very small compared to that, for example, of small-molecule solutions with near-ideal behavior.

3.14.2   Enthalpy of mixing

Dispersion of the polymer into the matrix results, as we have seen, in direct contact with the solvent molecules. Save for the excluded volume rule (no more than one molecule can exist per cell), no interactions were taken into account, suggesting no enthalpic changes during polymer solubilization. However, it is usual for polymers to have a non-zero enthalpy of mixing with solvent. In fact, during the coming together of polymer segments and solvent molecules, new bonds are formed. Let us consider that during mixing of solvent with polymer the interactions holding together the solvent molecules A–A and those holding together the polymer segments B–B are eliminated as to allow for new interactions between single molecules of solvent and molecules comprising the polymers. In chemical terms, this can be visualized with the following equation:

$$\frac{1}{2}\text{A-A} + \frac{1}{2}\text{B-B} \rightarrow \text{A-B}$$

The change in energy per new formed contact Δw_{AB} is called interchange energy and, according to the above stoichiometry, it is provided by the relationship

$$\Delta w_{AB} = w_{AB} - \frac{1}{2}(w_{AA} + w_{BB}) \tag{3.83}$$

where w_{AA}, w_{BB}, and w_{AB} are the enthalpic components associated with the terms in the previous chemical equation. Considering that no alteration in volume occurs during mixing, then for the establishment of F new contacts, the enthalpy of mixing ΔH^M is

$$\Delta H^M = F \Delta w_{AB} \tag{3.84}$$

Considering that each polymer segment neighbors with $z - 2$ cells (as two cells are occupied by its neighbors of the polymer itself), the entire chain has $r(z - 2)$ direct contacts with cells unoccupied by the same polymer, plus two more due to the fact that its terminal monomers connect with only one monomer. The number of external contacts per macromolecule f is $r(z - 2) + 2$. Assuming that r is a large number so that $r(z - 2) + 2 \approx r(z - 2)$ and that z is much greater than 2 so that $z - 2 \approx z$, we get the *Flory–Huggins approximation* $f = zr$. As per the discussion on the calculation of entropy, the possibility that a cell will be occupied by a solvent molecule equates to its volume fraction φ_1. Based on this approach, the number of the solvent molecules in direct contact with the polymer is $zr\varphi_1$. For N_2 polymer molecules, the enthalpy of mixing will be

$$\Delta H^M = N_2 rz\varphi_1 \Delta w_{AB} = N_1 z\varphi_2 \Delta w_{AB} \tag{3.85}$$

The interchange energy per solvent molecule will be $z\Delta w_{AB}$. This energy, expressed in units of kT, is called the *Flory–Huggins parameter* χ:

$$\chi = \frac{z\Delta w_{AB}}{kT} \tag{3.86}$$

We can rewrite Equation (3.85) using this very important physico-chemical parameter as

$$\Delta H^M = kT\chi N_1 \varphi_2 = RT\chi n_1 \varphi_2 \tag{3.87}$$

3.14.3   Gibbs free energy of mixing

We can combine the Equations (3.80) and (3.87) that we established for the entropic and enthalpic components ΔS^M and ΔH^M for the mixing of polymer and solvent to obtain the Gibbs free energy of mixing $\Delta G^M = \Delta H^M + T\Delta S^M$:

$$\Delta G^M = \Delta H^M - T\Delta S^M = kT(N_1 \ln\varphi_1 + N_2 \ln\varphi_2 + N_1 \varphi_2 \chi) \tag{3.88}$$

It is usual to express the free energy of mixing in terms of chemical potential for the solvent and μ_2 for the polymer. Remembering that the chemical potential μ is the molar free energy, Equation (3.80) can be used to obtain an expression for the chemical potential of the solvent:

$$\left(\frac{\partial G^M}{\partial N_1}\right)_{P,T,N_2} = \mu_1 - \mu_1^\circ = RT\left[\ln(1-\varphi_2) + \left(1-\frac{1}{r}\right)\varphi_2 + \chi\varphi_2^2\right] \tag{3.89}$$

Here, μ_1° stands for the chemical potential of the pure solvent. The activity of the solvent can be connected to the chemical potential:

$$\frac{\mu_1 - \mu_1^0}{RT} = \ln\alpha_1 = \ln(1-\phi_2) + \phi_2\left(\chi\phi_2 + 1 - \frac{1}{r}\right) \tag{3.90}$$

3.15   Osmotic pressure of solutions of macromolecules

Osmotic pressure plays a fundamental role in regulating the functionality and properties of biological systems. As we have seen in the discussion on small-molecule solutions, the osmotic pressure π of an ideal solution of molecular weight M is given by $\pi = RTm = c/M$, where m is the molarity ([mol substance] [dm^3 solution]$^{-1}$) of the dissolved substance and c is the g dm^{-1} of this substance in the same system. According to the Flory–Huggins theory (which is not further expanded in the present text), the osmotic pressure of solutions of macromolecules of molecular weight M_2 and of concentration c_2 is given by a virial-like equation of the form

$$\frac{\pi}{c_2} = RT\left[\frac{1}{M_2} + \frac{v_2^{sp2}}{v_1}\left(\frac{1}{2} - \chi\right)c_2 + v_2^{sp3}c_2^2 + \ldots\right] \tag{3.91}$$

where $v_2^{sp} = V_2/M_2$ is the specific partial volume of the polymer, which is usually written for simplicity as

$$\frac{\pi}{c_2} = \frac{RT}{M_n} + B_2c_2 + B_3c_2^2 + \ldots \tag{3.92}$$

$$\frac{\pi}{c_2} = RT\left(\frac{1}{M_n} + A_2c_2 + A_3c_2^2 + \ldots\right) \tag{3.93}$$

where $B_2 = RT\, A_2$.

3.15.1   The Donnan effect

A further contribution to the osmotic pressure of a polymer is the Donnan effect. It manifests at solutions of *polyelectrolytes*, that is, polymers that accommodate charged moieties. The Donnan effect is caused by the inherently unbalanced distribution of charges between a polyelectrolyte and its solution. Let us imagine a polyelectrolyte in solution, containing z negatively charged moieties (i.e., dissociated carboxyls). These charges are to be eliminated by an adsorbed layer, say of z [mol of protein] $= m_1$ mol of sodium counterions Na$^+$ (the notion of counterions is further discussed in Chapter 6). It is easy to calculate m_1 by multiplying its net charge by its molar concentration. Let us also imagine a membrane impermeable to large molecules (such as the polyelectrolyte) separating the previous solution A from a compartment containing water B, into which m_2 mol of NaCl are inserted. Complete dissociation of NaCl into m_2 mol of Na$^+$ and

m_2 mol of Cl^- will result in the development of an osmotic pressure due to the discrepancy in Na^+ and Cl^- electrolyte concentration between the two compartments (we will only briefly discuss here the osmotic effect due to the polyelectrolyte itself). According to Equation (3.45), excess concentration of a component, say Cl^-, between the two compartments will lead to spontaneous flux of Cl^- anions to the other compartment, followed by an equal amount of Na^+ cations to preserve the electroneutrality of the system. However, as the initial amount of Na^+ is bound to the polyelectrolyte (and has very restricted freedom, hence entropy: remember, osmotic pressure is an entropy-driven process), we may accept on the grounds of simplicity that it does not contribute directly to the overall osmotic pressure. According to the above, x mol of Cl^- and x mol of Na^+ are to be transferred into the polyelectrolyte, leaving behind $m_2 - x$ mol of Na^+ and $m_2 - x$ mol of Cl^- ions. The equilibrium condition for the osmotic equilibrium demands the activity product $a_{Na}^+ a_{Cl^-} = \gamma[Na^-]\gamma[Cl^-]$ to be equal for both compartments A and B. Assuming that γ is the same, we obtain $[Na^+]_A[Cl^-]_A = [Na^+]_B[Cl^-]_B$, which becomes the amount that moves from one compartment to the other:

$$x = \frac{m_2^2}{m_1 + 2m_2} \tag{3.94}$$

The overall effect of the above is that the small ions will increase in concentration on the side of the protein: As the concentration of the ions of Na^+ and Cl^- drops by x in the polymer-free compartment, the relevant concentration in the polymer compartment will be respectively increased.

The Donnan effect is of importance in proteins, which are charged when not at their pI. They can also be of importance in the case of polysaccharides, many of which bear charged groups. In the case of proteins, the regulation of the flow between the cellular and the extracellular fluids is largely determined by osmotic pressure, to which the Donnan effect is considered of very high importance.

3.16   *Concentrated polymer solutions*

The above section pertained to relatively dilute solutions. By "dilute" here we suggest solutions whose component macromolecules are at a reasonable distance from each other so that their movement is not significantly hindered. Let us imagine similar polymer molecules being added into the system so that the overall polymer concentration increases. As the concentration of polymers increases, the polymer molecules begin to influence each other.

Correlation length: The correlation length ξ is a measure of the range over which fluctuations in one part of the solution influence those in another part. If two infinitesimal parts of the solution are separated by a distance larger than the correlation length, they will each have fluctuations that are relatively independent of each other ("uncorrelated"). In very broad terms, the correlation length of a dilute solution has a value of ξ on the order of the size of the polymer.

Further addition of polymer will result in occupancy of most of the available space: From a given concentration onward, the new molecules cannot fit easily except into very specific vacant spaces in the system. This is called the *semidilute regime*, and is characterized by changes in the rheological behavior of the polymer solution. A characteristic of the semidilute regime is that the mass distribution presents maxima and minima, following an oscillating pattern with distance. This is because in a partially folded polymer, its center is denser than its outer regions. The wavelength of this oscillating pattern (i.e., the distance between two neighboring maxima) approximates the correlation length of the system. As the concentration increases further, there is no space to accommodate any more molecules, so the existing ones must intermingle with each other. If the system remains stable at such high polymer concentrations, it is considered as being in the *concentrated regime*. In this regime, the concentration of the polymer appears fairly even throughout the solution.

3.17   Phase separation

Solubilization of a component requires that the free energy of mixing ΔG^M should be lower than *any* possible combination of the components of the two co-existing phases. This means that for ΔG^M to be simply lower than the sum of free enthalpies of the pure components is not sufficient to ensure full mixing. Let us visualize the above in schematically drawn Figure 3.9. In the upper part the figure shows the dependence of the free energy of mixing to the molar fraction of the polymer X_{pol}. This part of the figure is, in essence, similar to the one depicted in Figure 3.3. At temperature T_3, at the top of the plot, solubilization occurs during incorporation of polymer into solvent and vice versa, as this leads to a reduction in ΔG^M until the minimum is attained, where $\partial \Delta G^M / \partial X_{pol} = 0$. After temperature T_c, known as the upper critical solution temperature (UCST), two minima are observed in the plots instead of one (i.e., T_1 and T_2). What happens in the case of a system with a composition between these minima? Such a solution is depicted as composition X_1 in Figure 3.9. This system will *phase separate* into two phases of compositions X_{1A} and X_{1B}. At T_c, the two minima merge into one.

As mentioned previously, in the case of a single minimum, there is only one point where the first differential $\partial \Delta G^M / \partial X_{pol}$ is equal to zero. This point is the minimum at T_3 and T_c. Multiple peaks such as those observed for T_1 and T_2 have both minima and inflection points. The inflections

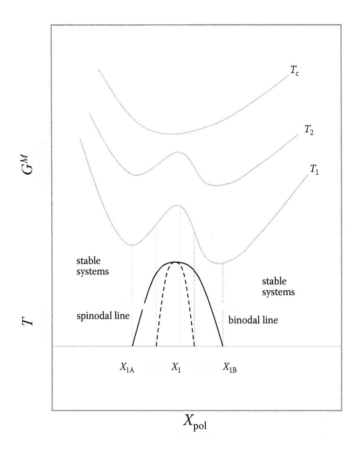

X_{pol}

Figure 3.9 Plots of the free energy of mixing (top) and of temperature (bottom) versus the molar fraction of a polymer in solution. Observe the dependence of the bimodal and spinodal lines, defining stability regimes, on the shape of the plots in the upper part of the plot.

observed should be associated with a zero value in the second differential of $\partial^2 \Delta G^M / \partial X_{pol}^2$. The first temperature in which the first, second, and third derivatives are zero is T_c. In terms of chemical potential, this can be expressed for T_c as

$$\frac{\partial \mu_1}{\partial \varphi_2} = \frac{\partial^2 \mu_1}{\partial \varphi_2^2} = 0 \tag{3.95}$$

This can be applied to Equation (3.90). The first derivative is

$$\frac{1}{1 - \varphi_{2c}} - \left(1 - \frac{1}{r}\right) - 2\varphi_{2c}\chi_c = 0 \tag{3.96}$$

where φ_{2c} is the critical polymer volume fraction and χ_c is the critical Flory–Huggins parameter. The second derivative is

$$\frac{1}{\left(1-\varphi_{2c}\right)^2} - 2\chi_c = 0 \tag{3.97}$$

These equations yield

$$\varphi_{2c} = \frac{1}{1+\sqrt{r}} \approx \frac{1}{\sqrt{r}} \tag{3.98}$$

The last simplification is justified by taking into account that r is very large for polymers. The critical Flory–Huggins parameter can be approximated as

$$\chi_c = \frac{1}{2}\left(1+\frac{1}{\sqrt{r}}\right)^2 = \frac{1}{2}+\frac{1}{\sqrt{r}}+\frac{1}{2r} \tag{3.99}$$

The lower part of Figure 3.9 is a plot of the molar fraction of the polymer with temperature. In accordance with the upper plot, three areas can be distinguished: an area outside of the curves, where stable polymer solutions are encountered; an area between the *binodal* and the *spinodal* lines, where metastable systems are normally formed; and the area enclosed by the spinodal line, where systems are expected to be inherently unstable and to phase separate, in a manner similar to that discussed for the solution of composition X_1.

3.17.1   Phase separation in two-solute systems

So far, the discussion has revolved around systems containing one type of polymer dissolved in a solvent. In reality, foods—perhaps the most complicated of all materials—contain a large number of macromolecules different to each other, normally in aqueous solutions and at relatively high ionic strength. As long as the macromolecules contained in a food do not interact strongly with each other, dilute solutions consisting of two or more kinds of soluble macromolecules can be stable. As their concentration increases, or as their conformation is expanded (i.e., via denaturation), their excluded volumes may start overlapping. This will decrease their freedom of movement, and hence decrease their conformational entropy. If this loss of entropy is not compensated by an equivalent decrease in enthalpy (i.e., by means of interactions with neighboring macromolecules), the system will undergo phase separation. Direct interactions readily

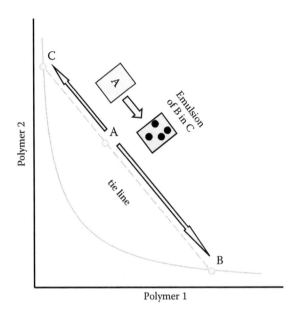

Figure 3.10 Schematic phase diagram for a solution of two polymers, depicting the tie-line and the phase separation of mixture A into an emulsion of solution B into solution C. In the case of aqueous solutions, this would be a water-in-water emulsion.

occur when the two macromolecules are similar. Their direct interaction is not unlike a solvation of one macromolecule into the other. When the two macromolecules are different in composition, however, and interactions cannot readily occur between them, *phase separation* can occur.

Figure 3.10 is a highly schematical depiction of what happens in a typical solvent containing two different macromolecular populations, referring to them as polymer 1 and polymer 2. This is a very usual scenario in food systems, where the solvent is water at a given pH and ionic strength/composition, and polymers 1 and 2 are proteins and/or carbohydrates.

EXERCISES

3.1 Estimate the molar fractions for the components of both the vapor and liquid phase for an ideal binary solution with components of vapor pressures 0.2 atm and 0.5 atm, assuming that the total vapor pressure is 0.35 atm.

Solution: Apply Raoult's law for binary mixtures; apply Equation (3.14).

Introduction to the physical chemistry of foods

3.2 The equilibrium constant of a reaction triples between 298°C and 328°C. Find the standard enthalpy of the reaction.

Solution: Van 't Hoff equation; define $\ln (K_2/K_1) = \ln 3$.

3.3 Measurements of the equilibrium constant K of the binding of an inhibitor of enzymic browning in vegetables have been carried out. The measurements were taken over a temperature range of 17.2°C to 38.2°C and are presented in the following table.

T (°C)	K 10^{-7}
17.2	7.1
22.0	5.6
25.0	6.2
28.2	3.9
32.1	2.9
38.2	1.9

Calculate the ΔG^0, ΔH^0, and ΔS^0 values of the binding of the inhibitor at 25°C.

Solution: Line plot $1/T - \ln K$; ΔH^0 from the Van 't Hoff equation; ΔG^0 from $\Delta G^0 = -RT \ln K$; and ΔS^0 from $\Delta G^0 = \Delta H^0 - T\Delta S^0$.

3.4 The equilibrium constant for a reaction is 670 mol L^{-1} at 25°C. Thermal measurements showed a change in enthalpy equal to –95 kJ mol^{-1}. Assuming that this is independent of temperature for the range of 10°C, calculate the reaction constant at 35°C.

Solution: Equation (3.34).

3.5 Calculate the ionic strength of
   a. A solution of 3% w/w NaCl
   b. A solution of 0.25 M $NaNO_3$
   c. A 3 mol L^{-1} solution of a protein with ten ionized carboxyl groups.
   d. The same solution at acidic pH (the protein has three ionized carboxyl groups).

Solution: Equation (3.54).

3.6 A weak acid has a K_a equal to 6.3×10^{-4}. Calculate the ratio of its conjugate acid and base at pH 3.

Solution: Calculate pK_a. Solve the Henderson–Hasselbalch equation for pH = 3.0.

3.7 A protein solution (0.2 mmol L^{-1}, pH 7) is placed into a compartment separated by means of a semipermeable membrane from a second compartment containing an equal volume of NaCl solution (75 mmol L^{-1}, pH7). Calculate the amounts of chloride and sodium ions in the two compartments after equilibrium has been restored. Consider that the protein has a net charge of −4.

Solution: Solve as a Donnan equilibrium using the relevant equation. Consider that z [mol of protein] = m_1.

chapter four

Surface activity

4.1   Surface tension

In solid objects, the atoms interact with each other with strong forces. These forces also exist, albeit to a lesser extent, in fluid systems. For example, the consistency of liquid water is mostly due to the existence of hydrogen bonds: Every molecule of water can bind with another four water molecules (two from the oxygen atom, and one each from the two hydrogen atoms).

Think about the physical boundary of a liquid system, for example the surface of water. Clearly, the molecules of water that are found in the last layer before the air cannot form bonds in all directions. Hydrogen bonds can be formed "downward" (toward the main body of water) but not "upward" (toward the air). Given that the physical significance of the bonds is to reduce the energy of a system, we can conceive that the inability of the surface water to form all the possible hydrogen bonds leads the molecules in question to a higher energetic state than the molecules in the body of the water. This appears macroscopically as additional free energy due to the existence of a free surface. This additional free energy in the surface is called the *surface tension* γ:

$$\gamma = \frac{\Delta G}{A} \tag{4.1}$$

The factor γ is directly related to the work that must be expended by a molecule in order for it to move from the continuous phase to the surface. This energy of all the molecules that move to the surface leads to the creation of an unstable system that, at the first opportunity, will try to minimize its surface area. A simple experiment that shows this tension uses an arrangement with a wire bent to form three sides of a rectangle, on top of the legs of which is balanced another straight wire in such a manner as to form the fourth side. The structure is dipped in a liquid and removed so as to leave a film held within the rectangle. If we slowly draw the free wire away from its opposite side of the square by a distance *l*, thereby increasing the area enclosed, then a force will be exerted on the wire that is equivalent and opposite to the force required to move it. For every liquid, this force will differ and will be proportional to the surface

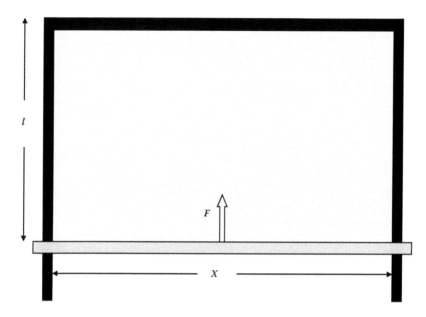

Figure 4.1 The experiment with the wire frame on which a free wire can be moved by distance *l*. The force exerted is proportional to the distance from the fixed frame, and the work of the movement is proportional to the area of the surface.

tension γ. Because the force is exerted along the whole length of the free wire, the surface tension will be

$$\gamma = \frac{F}{2l} = \frac{Fx}{2lx} = \frac{w}{A} \qquad (4.2)$$

where w is the work expended by the movement of the free wire and A is the area of the rectangular frame (Figure 4.1). The "2" in the denominator means that the forces are exerted on both sides of the liquid film.

For a change in the structure of a system of molecules of initial total pressure P and temperature T, the work w that is supplied is

$$w = P\Delta V + V\Delta P \qquad (4.3)$$

For the main bulk of a material, in other words the majority of its molecules, $\Delta P = 0$ under stable conditions. This does not apply to the surface, where with the application of a tangential force the pressure changes for the relatively small number of molecules that are found there. Thus, if we remember from the previous example with the wire that surface tension is the additional free energy (work) required to increase the surface by A, then we have

$$\Delta w = P\delta V + \gamma\delta A \qquad (4.4)$$

In this case, the distance between the surface molecules will increase slightly (let us say by z) and, at the surface only, the tangential pressure will become P_T. The surface tension will then be

$$\gamma = \int (P - P_T)dz \qquad (4.5)$$

The existence of this free energy drives substances to organize themselves in order to minimize their surface area-to-volume ratios. The geometric shape that presents the smallest such ratio is the sphere. An immediate consequence of this is that substances condense where possible to form spherical globules (e.g., droplets and bubbles) as long as this is permitted by other potentially interfering forces such as gravity. In reality, as shown by Laplace and Young, the geometric shape of droplets such as those formed on top of a surface and menisci such as those formed in capillary tubes are characterized by two radii of curvature (R_1 and R_2), rather than just one as in a simple sphere. These are connected to the surface pressure ΔP by the Young–Laplace equation.

$$\Delta P = \gamma \left(\frac{1}{R_1} + \frac{1}{R_2} \right) \qquad (4.6)$$

By correlation, *line tension* can be defined as the energy per unit length of the edge of a space that contains a substance. Line tension is expected to be larger than surface tension because the molecules that are found along the length of a line can form still fewer bonds, such as hydrogen bonds, with neighboring molecules. Line tension is observed at the edges of geometric shapes, for example, at the point of contact of a droplet with a surface.

The shape of a droplet on a surface is spherical due to the attempt of the droplet to minimize the free surface area and because of the symmetrical distribution of forces. This is due to, among other things, the capillary phenomenon, as we will see later. The corresponding *point tension* has been described and studied theoretically.

4.2 Interface tension

As mentioned in the previous section, surface tension is the additional free energy per unit surface that results from the inability of the surface molecules to form interactions with other molecules in the direction of the empty space above the surface. If we bring two substances into contact in order to form two separate phases that are separated by an *interface*, then maybe the molecules of the two phases that are found at the interface can exert some forces between them. These forces may be weaker than those

exerted between molecules of the same phase, but they can reduce somewhat the free surface energy. Thus, if we bring into contact two substances
A and B with surface tensions γ_A and γ_B, respectively, then the interface
tension γ_{AB} will be

$$\gamma_{AB} = \gamma_A + \gamma_B - 2\sigma_{AB} \tag{4.7}$$

where σ_{AB} is the energy gain per surface unit area due to attractive interactions between the molecules of A and B. This equation shows that the free
energy per unit area γ_{AB} is large when the surface tensions are large and
the interactive forces between the phases are small.

As each of the two populations of surface molecules A and B tends
to retire to its respective bulk phase A or B, the two surfaces tend to dissociate from each other. The sum $\gamma_A + \gamma_B$ quantifies the energy per unit
area relevant to this dissociation. These two surfaces are mutually *insoluble* and will form a clear interface between them. At the macroscopic
level, the two components A and B will *phase separate*. If one considers a
second scenario where the forces between A and B increase and become
comparable to the forces leading to phase separation, then Equation (4.7)
becomes $\gamma_A + \gamma_B = 2\sigma_{AB}$. In this case, γ_{AB} is zero and the surface energy
is also zero. This will lead to conditions in which the two phases are
thermodynamically compatible, the interface will break down, and complete *mixing* will occur.

4.2.1   A special extended case

Let us think of a layer of molecules of a third substance C between two
immiscible liquids A and B. The molecules of C have one end that can
interact strongly with phase A, and the other with phase B. The layer of
the third substance C, being related with both substances A and B, will
transform Equation (4.7) into $\gamma_{AB} = \gamma_A + \gamma_B - (\sigma_{AC} + \sigma_{BC})$, with the values of
σ comparable in size to the values of γ. The two incompatible liquids A
and B can now be rendered miscible with the mediation of the layer of
substance C, which is called a *surfactant*.

When the two substances can interact with similar forces, Equation (4.7)
becomes

$$\gamma_{AB} = \gamma_A + \gamma_B - f(\gamma_A\gamma_B) \tag{4.8}$$

The general equation that gives the change in the total energy of an
interface during its expansion or contraction is

$$U = TS + \gamma A + \sum_i \mu_i n_i \Rightarrow dU = TdS + \gamma dA + \sum_i \mu_i dn_i \tag{4.9}$$

where S is the entropy of the system, μ_i the chemical potential of component i, and n_i^σ its additional mole fraction in relation to the corresponding concentration in the continuous phase.

4.3 Geometry of the liquid surface: Capillary effects

The decisive contribution of surface tension to the shape of a liquid droplet was discussed previously. Let us come back to that example and consider a droplet on a solid surface (Figure 4.2). Two categories of forces apply to the molecules that are found in the ring of liquid that is in tangential contact with the surface: (1) *cohesive* forces and (2) *affinity* forces. Cohesive forces concern the forces that develop between the molecules of the liquid, while affinity forces develop between the molecules of the liquid and those of the solid surface at the interface. The combined action of the two forces determines the *contact angle* θ that is formed by the tangent of the meniscus and the surface. The contact angle depends on the thermodynamic adhesion between the liquid and solid. If the cohesive forces are greater than the affinity forces—that is, the miscibility of the molecules of the solid and liquid is relatively low—then the angle θ is obtuse. The final value of the contact angle can be determined by the values of three interface tensions: That of the air-liquid, that of the solid-air, and that of the solid-liquid interface (see Figure 4.2). This is the case with, for example, a drop of water on a hydrophobic surface such as PTFE. If the adhesive forces are greater than the cohesive forces, that is, the miscibility of the

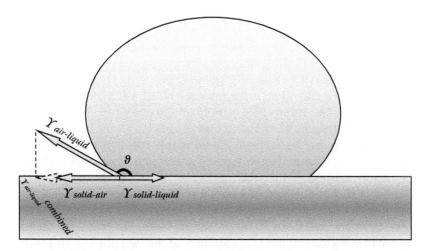

Figure 4.2 Schematic representation of a droplet on a surface. The contact angle θ is determined by the resultant of the tensions of the individual interfaces.

molecules of the solid and liquid is high, the angle θ is acute. This is the case, for example, for a drop of water on a hydrophilic wooden surface.

Corresponding phenomena occur in the case of liquids in tubes. Careful observation of a glass tube containing water will often reveal that the water at the walls of the tube is at a higher level than that in the center. More revealing is when one end of a very thin tube (capillary) is immersed in a container of water, the level of the water in the tube rises higher than the level in the main vessel, an apparent violation of the principles that apply to interconnecting vessels. Conversely, if the tube is made from hydrophobic polymer, the water in the tube is lower than that in the main container. In addition, the water in the capillary tube forms a meniscus. The above phenomenon is called *capillary action* and again connects the cohesive forces of the liquid and the adhesive forces between the liquid and the capillary tube. As long as there is a good thermodynamic relationship between the liquid and the material of the tube, it is possible for the adhesive forces to overcome not only the cohesive forces but also gravity, so that the liquid rises up the tube. On the contrary, when the thermodynamic affinity between the liquid and the tube is not strong, the cohesive forces can overcome the adhesive and gravitational forces, pulling the liquid in the tube down below the level of the liquid in the main container.

4.4   Definition of the interface

The existence of an interface between two substances drastically changes the thermodynamic status of a two-component system. As previously mentioned, this is due to the fact that the interface has different energetic properties from the main bulk of the system. That means that, moving from the bulk of component A and passing through the interface into the bulk of component B, the properties of the system change from those of pure component A to those of pure component B. The area where this transition comes into effect is the interface. The basic problem in describing a system of two immiscible components is describing how a property changes with distance from the interface.

The most successful basis for tackling this problem was proposed by Gibbs: In a two-component system of immiscible liquids (A and B), the interface is defined as the level of contact of the two liquids. The basic premise is that the values of a property (e.g., concentration of a solute, conductivity), is constant throughout the bulk of the continuous phase, as suggested previously, but changes from the value in A to the value in B in a narrow region around the boundary of the two phases. This narrow region that is inserted between the two immiscible liquids is called the *Gibbs surface* (see Figue 4.3).

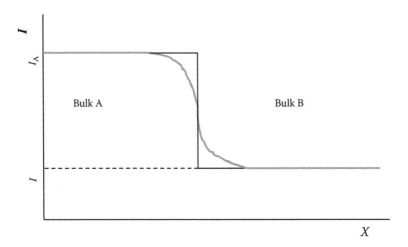

Figure 4.3 Change in property I as a function of the distance from the interface between two surfaces A and B. Throughout the distance Δx, the value of the property changes progressively from I_A to I_B. In contrast, the thin line shows the sudden change in the property under Gibbs' model.

As mentioned, the basic feature of the interface according to Gibbs is that a property changes suddenly from phase A to phase B. In reality, there is a region several molecules thick with physical and chemical properties that are different from both phases A and B. If we try to form an image of this interface in our minds, we have to envisage it as a region with its own properties in which the transition from the main bulk of phase A to the main bulk of phase B is regular and continuous. Usually (but with many exceptions) properties such as the concentration of adsorbed substances change smoothly in a distance of 3 to 5 molecular diameters.

The main disadvantage of Gibbs' approach lies in the fact that, with the exception of a small area close to the separating surface, the properties are considered unchanging as a function of distance. Even if for properties such as the concentration of surfactants this is generally correct, it is wrong to consider that there is a real dividing line between the continuous phases and the interface, or that the Gibbs interface has any physical status. Gibbs' approach, however, continues to be the teaching standard and the basic model for examining the properties of systems of two immiscible phases.

4.5   Surface activity

In the preceding paragraph we followed the development of a property as a function of its distance from the interface between two phases.

Equation (4.9) calculates the value of the energy of an interface of i components. For the total energy, we have

$$SdT + Ad\gamma + \sum_i n_i d\mu_i = 0 \tag{4.10}$$

If we differentiate Equation (4.10), subtract it from Equation (4.9), and solve for $d\gamma$, we obtain

$$-d\gamma = \sum_i \frac{n_i^\sigma}{A} d\mu_i = \sum_i \Gamma d\mu_i \tag{4.11}$$

where Γ is the additional surface concentration of component i in mol m^{-2} beyond the stable concentration of component i in the main bulk of the continuous phase. Equation (4.11) is called the *Gibbs isotherm* and is one of the most important equations in surface science. It is one of those rare equations where a structural property on a scale of Angstroms (here the surface area of a single molecule, calculated from the surface area of a mol) can be calculated using macroscopic observations (measurement of surface tension). Its main application is the calculation of the *adsorption* of substances to surfaces. We can define adsorption as the increase in concentration of a substance in the region of the interface relative to the concentration values that are found in the continuous phase of a system. To study the adsorption of substance Q on the surface of material A, if we choose a point on the Gibbs interface such that $\Gamma_A = 0$, then we obtain

$$\Gamma_Q = \frac{-d\gamma}{d\mu_Q} \tag{4.12}$$

Because

$$\mu_Q = \mu_Q^0 + RT \ln x_Q \tag{4.13}$$

where x_Q is the mole fraction of substance Q, μ_Q^0 is the chemical potential under ideal conditions and R is the universal gas constant, we obtain

$$\Gamma_Q = \frac{-1}{RT} \frac{d\gamma}{d \ln x_Q} \tag{4.14}$$

Equation (4.14), sometimes expressed in terms of activity or concentration rather than mole fraction, tells us that the change in surface tension will be negative following an increase in the adsorbed substance, so the final surface tension will be *smaller than the initial* following the adsorption of a substance *at the interface*. If the adsorption of a substance increases the interface tension, then, according to the Gibbs equation, this substance will have a lower concentration at the interface than in the continuous phase. This is true for many inorganic salts dissolved in water. The above phenomena are discussed more extensively in Section 4.6.

4.6   Adsorption

The phenomenon of adsorption is a physicochemical peculiarity that is directly related to the existence of surfaces and interfaces. During the adsorption process, molecules are transferred from the main bulk of the system to the surface/interface. *Ad*sorption must not be confused with the more complicated concept of *ab*sorption, which relates to the transfer of a molecule from the main mass of one phase to the main mass of another. Adsorption, driving the molecules to the spatially limited area of the interface, increases their local concentration. For this reason, nature and man use this process, among others, in order to accelerate chemical reactions (such as on catalysts and enzymes), to form structures (such as the adsorbed interfacial layers of emulsions) or to separate a component from a mixture (as in chromatography and on filters).

4.6.1   Thermodynamic basis of adsorption

A particle in the main phase of a material (e.g., a molecule of gas or a substance dissolved in water) can be considered to move without hindrance in three dimensions. Once adsorbed, the particle loses free movement on the axis perpendicular to the surface, as such movement would equate to desorption (the opposite of adsorption). Movement of the particle is limited to the plane of the surface to which it is adsorbed. This limitation of movement from three to two dimensions constitutes a reduction in entropy. From the second law $\Delta G = \Delta H - T\Delta S$ it is evident that for a counter-entropic phenomenon such as adsorption to take place spontaneously, the enthalpy must be reduced, so adsorption must in principle be exothermic. Because of this, under equilibrium conditions, adsorption is more extensive at low temperatures and is limited at high temperatures.

4.6.2   Adsorption isotherms

Let us consider molecules of gas adsorbing onto a solid surface or solute molecules adsorbing onto an interface. As we have seen, adsorption involves the transfer of molecules from the continuous phase (here the gas or solvent) to the surface/interface. The process can be presented as the reaction $X_{bulk} \rightarrow X_{ads}$, during which the molecules from the main mass of gas or liquid (X_{bulk}) adsorb onto the interface (X_{ads}). As with every reaction, Le Chatelier's principle applies, that is, the quantity of adsorbed substance *q* (usually but not always expressed as mass or molecules per adsorbent mass) will increase with the concentration *C* or pressure *P* of the molecules in the gas phase.

Adsorption data under constant temperature is usually presented as plots of *C–q* or *P–q*, while for gases the form *P-V* (where *V* is the volume of the adsorbed gas) is common. These plots are called *adsorption isotherms*. As mentioned in Section 4.6.1, adsorption is directly related to temperature. In cases where the temperature is not constant, the data can be presented as *isobaric* or *isosteric* for adsorption under constant pressure or constant amount of gas, respectively.

Perhaps the most simple adsorption isotherm is given by Henry's equation, according to which the surface concentration of the adsorbing substance *q* is proportional to its concentration *C* (for solutions) or its pressure *P* (for gases) in the continuous phase:

$$q = KC \qquad\qquad (4.15)$$

$$q = KP \qquad\qquad (4.16)$$

Henry's law of adsorption is basically similar to his law for the absorption of a volume V_{abs} of gas by solid bodies:

$$V_{abs} = KP \qquad\qquad (4.17)$$

Henry's law can apply to very small pressures or concentrations on homogenous surfaces, as long as the components of the adsorbed material do not interact with each other. Henry's equation must be modified for systems such as those we encounter in foods. Some of the first satisfactory modifications were carried out in the first half of the twentieth century.

The first thorough mathematical description of adsorption under isothermal conditions was given by Langmuir with the equation that bears his name:

$$q = \frac{QbC}{1+bC} \qquad\qquad (4.18)$$

In the Langmuir isotherm, *b* represents the energy of adsorption, while *Q* is the greatest possible surface concentration (expressed in a similar way to *q*).

Isotherm (4.18) can be rewritten based on the ratio *q/Q*, namely the ratio of the actual surface concentration of the adsorbed substance to the maximum possible concentration:

$$\frac{q}{Q} = \frac{bC}{1+bC} \qquad\qquad (4.19)$$

For a gas, the ratio q/Q can be written in terms of pressure and volume. More particularly, the first part of Equation (4.19) equates, under normal conditions, to the ratio of the volume of adsorbed gas V to the total volume that can be adsorbed V_m. Furthermore, for gases under normal conditions, the concentration is proportional to the pressure P. Thus, Equation (4.19) can be rewritten as

$$\frac{V}{V_m} = \frac{BP}{1+BP} \Rightarrow \frac{P}{V} = \frac{1}{BV_m} + \frac{P}{V_m} \tag{4.20}$$

Equation (4.18) was produced with the assumption that the adsorption takes place at particular centers, each of which can hold one adsorbed molecule. Furthermore, the heat of adsorption (an enthalpic quantity) is considered to be the same for all the centers and independent of molecules that are already adsorbed. The adsorbed molecules are assumed not to interact with one another. Although the Langmuir isotherm can satisfactorily describe many phenomena, it fails to describe systems with heterogeneous surface morphology or cases with multiple adsorbed surface layers. These weaknesses drove the development of other equations such as the empirical equation of Freundlich:

$$q = Kc^{\frac{1}{n}} \tag{4.21}$$

In the Freundlich equation, the parameter K is a constant that relates to the adsorptive ability of the material, while the coefficient n in the superscript relates to the heterogeneity of the adsorbent surface.

Brunauer, Emmet, and Teller applied themselves to extending the power of the Langmuir isotherm and developed an adsorption isotherm that is known by their initials (BET). The BET isotherm is based on the concept of successive adsorption of consecutive layers of molecules of a gas on a surface and describes the serial adsorption of gas molecules (1) on a free surface (resulting in the creation of a new surface layer), (2) additional molecules on top of the first adsorbed layer (resulting in the creation of a second layer), (3) additional molecules on top of the second layer (resulting in the creation of a third layer), and so on.

Considering the process of successive adsorptions as a sequence of reactions $X_{bulk} \rightarrow X_{ads}$ (an assumption we already made above), we can distinguish two subcategories of adsorbed layers: the first molecular layer that forms directly on the interface with greater energy of adsorption E_a, and the succeeding layers that adsorb more weakly onto molecules similar to themselves with a smaller adsorption energy E_c. The difference $E_a - E_c$ is called the *pure heat of adsorption*, while the equilibrium constant C for the adsorption of multiple molecules can be written

$$C = ge^{\frac{E_a - E_c}{RT}} \tag{4.22}$$

where g is a constant, R the universal gas constant, and T the temperature. If we consider that P_o is the vapor pressure of the adsorbed substance, V_m is the maximum volume that can be adsorbed as a monolayer, and C is the equilibrium constant from Equation (4.22) for volume V of gas adsorbed at pressure P, then the final form of the BET equation is

$$\frac{P}{V(P_o - P)} = \frac{1}{CV_m} + \frac{(C-1)}{CV_m} \frac{P}{P_o} \tag{4.23}$$

The importance of the BET equation lies in the fact that, based on experiments of adsorption of gas to a material, it is possible to calculate the specific surface area of the material in question. For this purpose, the adsorption of gas to a surface is studied in order to plot a diagram of

$$\frac{P}{V(P_o - P)} - \frac{P}{P_o}$$

According to Equation (4.23), the plot can be adjusted to a straight-line slope of

$$S = \frac{(C-1)}{CV_m} \tag{4.24}$$

while the intercept is, in principle,

$$I = \frac{1}{CV_m} \tag{4.25}$$

From these two equations we can calculate the volume V_m (the volume of gas required for single layer coverage):

$$V_m = \frac{1}{I + S} \tag{4.26}$$

Knowing V_m and substituting it into the ideal gas equation, we can calculate the number of adsorbed molecules. If A_o is the surface area that is occupied by a single isolated adsorbed molecule and V_{mol} is the molar volume of the gas, the total surface of the solid A_σ can be calculated from the equation

$$A_\sigma = \frac{V_m N_A}{V_{mol}} A_o \qquad (4.27)$$

where N_A is Avogadro's number. With the above reasoning, the BET equation is widely used in the processing of results of adsorption experiments for the purpose of determining the porosity of materials. Porosity is of great interest in technologies such as gas traps, filters, catalysts, and also biomaterials (such as artificial bone replacements). As with the Gibbs equation, the reader should pay attention to the subtle and elegant way with which interfacial methods can provide us with the dimensions of molecules based solely on macroscopic observations.

Of particular importance in the field of food science is the equation of Guggenheim, Anderson, and de Boer, known from their initials as the GAB equation. This equation correlates the quantity of adsorbed moisture w in a material with the water activity a_w:

$$w = \frac{w_m C K a_w}{(1 - K a_w)(1 - K a_w + C K a_w)} \qquad (4.28)$$

In the GAB equation, the parameter w_m relates to the moisture that is required for the formation of a monomolecular layer on the surface of the adsorbent material (in general terms, this correlates to the parameter Q in the Langmuir equation), while the values C and K are constants that depend on the temperature, among other things.

The method usually preferred for the calculation of the parameters w_m, C, and K is the elaboration of the isotherms $a_w/w - a_w$ and the fitting of the graph to the equation

$$\frac{a_w}{w} = a + b a_w + c a_w^2 \qquad (4.29)$$

where

$$a = \frac{1}{w_m C K} \qquad (4.30)$$

$$b = \frac{C - 2}{w_m C} \qquad (4.31)$$

$$c = \frac{K(1 - C)}{w_m C} \qquad (4.32)$$

The coefficients w_m, C, and K are calculated from the secondary equation

$$aK^2 + bK + c = 0 \tag{4.33}$$

and its solutions

$$C = \frac{b}{aK} + 2 \tag{4.34}$$

$$w_m = \frac{1}{b + 2Ka} \tag{4.35}$$

The GAB equation can show discrepancies for very high water activities, and for this reason several modifications have been proposed.

4.7 Surfactants

As mentioned previously, the basic use of the Gibbs isotherm (Equation (4.14)) is the calculation of the surface concentration from the measurement of surface tension. The reduction in surface tension is due to the adsorption to the interface of substances at surface concentrations of Γ above that in the continuous phase. Which substances will adsorb selectively from a solution? Or to put it another way, which substances are those that satisfy the condition of Equation (4.14) for reduction in γ with a smaller value of Γ and x_Q? Why do some particular substances that are adsorbed at a lower concentration on a surface from a lower concentration in the continuous phase cause a greater reduction in interface tension? We saw previously in (4.21) the case in which a third compound can be interpolated in a monomolecular layer into the interface between two substances A and B, and this can lead to the two phases becoming thermodynamically compatible (γ_{AB} becomes 0). Such a compound would reduce the interface tension and, according to the Gibbs equation, adsorb selectively onto the interface.

We have described the basic prerequisite that such a molecule must fulfill: It must have parts that are soluble in both phases A and B. If we extend this line of reasoning and say that the two phases are immiscible, then we can describe a surfactant molecule as a molecule that has two parts, one that is lyophilic in one system and lyophobic in the other and vice versa. In most cases, two distinct phases A and B are usually immiscible due to their differences in polarity. In a typical food scenario, one phase can be said to be more polar than the other; that is, one phase is a continuous aqueous phase, while the other is a continuous triglyceride phase ("fat"). Any molecule that has distinct polar and nonpolar parts can be called a surfactant. For aqueous systems, a surfactant can be any molecule that has distinct hydrophilic and hydrophobic parts. A consequence of the presence of both lyophilic and lyophobic parts in the same molecule is the fact that there will be thermodynamic complications with the dissolution of a surfactant in any single bulk phase while, inversely, there will always be thermodynamic incentive for a surfactant to adsorb selectively onto interfaces and thus to reduce the surface tension.

In nature, colloidal dispersions of some micrometers in diameter and a combined surface area of 10 m² mL⁻¹ may be encountered. From a thermodynamics point of view, such systems should not exist due to the large amount of energy that is trapped in the interfaces, but they are possible with the intervention of surfactants. These adsorb to the interfaces and reduce the surface tension, and consequently the work required to form the free surface. The manner in which surfactants achieve this is discussed more extensively in the chapter on emulsions and foams (Chapter 6).

Surfactants accumulate on the interface and orientate themselves such that their polar ends are solubilized in the polar phase and the nonpolar parts in the nonpolar phase (Figure 4.4). In the case of aqueous systems (we persist with systems in which one of the two phases is water, as these are the most common), it is usually accepted that, due to the strong polarity of water, the polar parts are water soluble in contrast to the usually occurring nonpolar or less polar oil-soluble region. Because in most technological applications the nonpolar parts are long hydrocarbon chains, the hydrophobic parts of surfactant molecules tend to be called *tails*, while the usually (but not always!) smaller hydrophilic parts tend to be called *heads*.

According to the most basic Equation (4.14), we see that substances that reduce the surface tension accumulate on the surface. To the contrary, substances that increase the surface tension are found on the interface in lower concentrations than they are found in the continuous phase. We

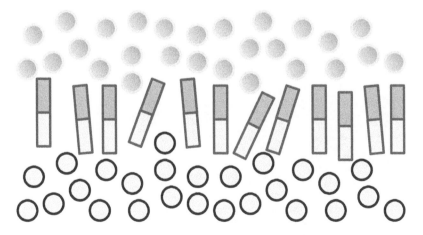

Figure 4.4 Schematic representation of the interface of substances A and B with the interpolation of a layer of surfactant molecules.

have seen previously (Section 4.5) that this can apply to some inorganic salts. What does this mean from the point of view of physical chemistry? The small ions of salts are more polar than water. Their presence makes the two phases even more incompatible, as the molecules of water are more soluble in the nonpolar phase than these salts. Thus, the salts are repelled from the interface into the interior of the aqueous phase.

chapter five

Surface-active materials

5.1   What are they, and where are they found?

The science of foods, decontaminants, coatings, foaming agents, and more generally of colloids, the general category to which many of the above belong, differs significantly from the classical physics and chemistry of solutions. This is largely due to the particular chemical nature of surfactant molecules. The latter are composed of two separate parts, one of which is polar and the other nonpolar. In a biphasic system, the surfactants position themselves on the interface between the two phases, with the most polar parts oriented (solvated) in the more polar phase and the nonpolar parts solvated in the less polar phase (in accordance with the old saying in chemistry that "like dissolves like").

Although the aim of the present work is to present the subject of surfactants from a practical point of view and with an emphasis on food chemistry, it is necessary for the theoretical basis of the subject to be presented in parallel. This has already been done in the previous chapter for thermodynamics, the physical chemistry of solutions, and the science of surfaces and interfaces, while the subject of the self-assembly of materials is covered in this chapter. It is the personal opinion of the author of the present work that a book concerning the science and technology of surfactants and colloids must not be limited to classical interface chemistry. New knowledge about the nature and organization of material reveals that the seemingly continuous phases are a macrocosm of interfaces and individual microphases and nanophases, a realization with huge practical consequences: The chemistry of biological systems has as its basis the chemistry of interfaces. Over the past three decades, additional steps forward in research into the organization of surfactants have shown that the seemingly anti-entropic processes of self-assembly are among the most important properties of nature and are probably the key to understanding not only the action of an emulsifier or a dye, but also the formation of life itself, the functioning of individual biochemical processes, and other broader philosophical questions.

5.2   Micelles

The importance of hydrophobic interactions in the physicochemical behavior of surfactant substances was stressed in preceding chapters. Briefly, the distancing of a hydrophobic molecule or group from an aqueous solvent leads to a reduction in the free energy of a system. This happens because a hydrophobic unit, such as a hydrocarbon, prefers to be in a nonpolar environment (e.g., with other hydrocarbon molecules) rather than a polar environment (such as water). When a hydrocarbon is removed from an aqueous matrix, water molecules that used to surround it form additional hydrogen bonds between them; in addition, the so-called hydrophobic attractions between adjacent hydrocarbon chains keep them close together and increase the incentive for phase separation between polar water and nonpolar hydrocarbons or similar molecules. In a biphasic system with a nonpolar (hydrophobic) phase and a polar (hydrophilic phase), the obvious course of action for a hydrocarbon that finds itself in the hydrophilic phase is to migrate into the hydrophobic phase and dissolve. However, in the absence of a hydrophobic phase, the best option for the hydrocarbons is to aggregate together, self-assembling into super-molecular entities in order to reduce contact with the water to a minimum (i.e., only the exterior of the supermolecular structures that are formed). These molecular structures can be considered a new hydrophobic phase in which further hydrocarbons can dissolve.

From a thermodynamic standpoint, the aggregation of hydrocarbons leads to a *self-assembled form* that ensures numerical equality between the energetic benefit from the formation of hydrophobic bonds (enthalpic factor) and the preservation of the greatest degree of freedom at a particular temperature (entropic factor). If the forces involved are considered, the final *form* of the matrix of self-assembling hydrocarbons is a result of the balance of attractive and repulsive forces, as explained later.

The phenomenon of self-assembly is very important for surfactants. The structures that are formed by amphiphilic (hydrophilic and hydrophobic regions together on the same molecule) surfactant molecules are directly related to the geometric structure of the molecules, their concentration, the temperature, and the nature of the solvent. The simplest form of aggregate in an aqueous medium is the *micelle*. This is a spherical body in which the nonpolar (hydrophobic) parts of the molecules congregate in the interior, while the polar (hydrophilic) heads orient themselves toward the outside and collect on the external surface of the micelle. Thus, the externally hydrophilic spherical aggregate that is formed is solvated in the aqueous environment.

Micelles (similar to almost all aggregates of small surfactant molecules) are dynamic entities. Generally, micelles should be considered, with the exception of two- and three-membered aggregates,[*] the simplest supermolecular structure of surfactants. In aqueous systems, their radii are comparable with the length of the hydrophobic group and for hydrophobic hydrocarbon chains of six to eighteen carbon atoms the size should be broadly on the order of a few nanometers. If they were smaller, the hydrophobic tails could be forced out of the hydrophobic core to the hydrophilic surface, which is not thermodynamically desirable. If they were larger, an undesirable void would be created between the end of the molecule and the center of the sphere.[†] A very large number of experimental measurements have verified first the dimensions of micelles, and second the homogeneity of their interior. From a physical point of view, the interior of a micelle can be considered a nonpolar solvent in which the hydrophobic parts of the molecules comprise The interior of the micelle which must be regarded as a liquid system comprised of lyophilic, flexible chains. In this way we can explain the reduction in the concentration of organic substances that dissolve in water—they are solvated in and pass into the interior of the micelles.[‡] If the surfactant is ionic, a very large percentage of the counter-ions can be found close to the surface of the micelle, and a state of dynamic equilibrium exists between counter-ions close to the interface and those in the continuous phase. This is expanded upon in Section 6.3.2.

A micelle can be regarded as the product of an exothermic condensation reaction of N surfactant molecules. The number N is called the *aggregation number* and for the same substance depends on the solvent, the temperature, and the ionic environment. The condensation reaction can be written as follows:

$$A_n + A \rightleftharpoons A_{n+1}$$

with equilibrium constants k_1 (to the right) and k_2 (to the left). These two constants are not the same as each other: k_1 depends on the migration of the surfactant from the continuous phase into the micelle and k_2 depends on the characteristics of the micelle itself (surface pressure, chemical characteristics of the molecules, temperature). Even after the formation of abundant micelles from monomers and oligomers, the exchange of

---

[*] Two or more molecules whose hydrophobic parts group together by means of hydrophobic interactions.

[†] In reality, the void in question can exist and is filled by other nonpolar molecules. In this case, the hydrophobic tails function as solvents for these molecules. Maybe this gives an indication of the mode of action of detergents?)

[‡] Continuing the theme of the previous footnote, when the accumulation of substances in the interior of the micelle is sufficient to be considered a clearly defined nonpolar phase, then this system of micelles is called a *microemulsion*.

molecules between dissolved monomers (A) and the surfactants of the micelle (A_n) continues to take place in exchange the order of milliseconds, while the process slows with the increasing size of the hydrophobic group. The reader must imagine a process of continual exchange of molecules between the micelle and the dissolved phase.

A micelle can be comprised of several tens of molecules. For the formation of a self-assembling structure such as the micelle, two forces are required: (1) one that promotes aggregation into particular structures and (2) another that facilitates their break-up into smaller units. It has already been mentioned that the driving force behind micelle formation is the hydrophobic interactions or, more generally, lyophobic interactions (which can include hydrophilic interactions in the case of inverse micelles as we will see below) between the lyophobic regions of the molecules. The force that counteracts the formation of micelles is the sum of electrostatic repulsions between ionic parts and of stereochemical repulsions between the hydrophilic heads of complex molecules (such as the nonpolar polyoxyethylene surfactants) and between the tails.

In the first stages of the formation of a micelle (when the number of molecules is clearly smaller than the aggregation number), the forces driving the formation of micelles dominate those opposing the aggregation. When the number of monomers approaches or tends toward exceeding the aggregation number, the forces governing aggregation and disintegration are generally equal. Thus, the micelle stabilizes in this state of dynamic equilibrium. This does not mean that the micelle is a static entity, but rather in a fluid state with molecules continually being exchanged between the micelle and the continuous phase.

5.3   *Hydrophilic-lipophilic balance (HLB), critical micelle concentration (cmc), and Krafft point*

In general, the selection of the right emulsifier, detergent, or surfactant for every application requires prior knowledge of some basic parameters. The most important technological parameters for surfactants in general are probably the values HLB, cmc, and the Krafft point.

The hydrophilic-lipophilic (hydrophobic) balance value (HLB) is exceptionally useful and constitutes a general guide to the use of a surfactant based on its hydrophilicity and hydrophobicity. Many different ways of calculating the HLB have been proposed (indicative are the methods of Griffin and of Davies), with the calculation of the total number or mass of the hydrophobic and hydrophilic regions of the surfactant forming the common ground between them. In general, an HLB value of 0 implies a completely hydrophobic molecule, while higher values (above 10) indicate hydrophilic surfactants. Hydrophobic surfactants are potentially good emulsifiers for water-in-oil emulsions (where water droplets are dispersed in oil), while hydrophilic emulsifiers are more suitable for oil-in-water emulsions (where oil droplets are dispersed in water). Practical rules for the use of surfactants can be found in the literature, although the stated limits can vary somewhat.

In the majority of cases, micelles are created spontaneously when the concentration of the surfactant passes a certain concentration known as the *critical micelle concentration* (cmc) (Figure 5.1). At this concentration, a sudden change in the macroscopic parameters of a surfactant solution is observed (among other things, the stabilization of surface tension and osmotic pressure, an increase in turbidity and scattering of light, an increase in magnetic resonance, and a reduction in conductivity). The changes are due to the sudden phase change from isolated molecules to micelles. The cmc is instrumental as a basic criterion for the evaluation of its relative effectiveness. Most of the physical parameters previously referred to as changing at the cmc can be used for its determination. The most typical methods are the measurement of surface tension, conductivity, and turbidity, but there are several other techniques, such as NMR (nuclear magnetic resonance) spectroscopy and fluorimetry. Equivalent techniques are used in defining other parameters, such as the critical concentration of other structures like vesicles, but their study lies outside the scope of the current work.

The cmc is perhaps the most studied property of surfactant solutions. Its value for a given category of surfactant depends on the length of the hydrophobic part of the molecule (which, as previously mentioned, usually consists of one or more carbon chains), the chemical structure of the hydrophobic part (any substituent groups or double/triple bonds can play an important role), the charge and size of the hydrophilic part (particularly in polyoxyethelyne surfactants), and the temperature and nature of the solvent (presence of salts, nature of counter-ions, nonaqueous solvent).

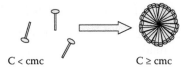

C < cmc   C ≥ cmc

Figure 5.1 Micelles are spontaneously created when the concentration of individual surfactant molecules exceeds the cmc.

The effect of structure of the molecule on cmc is easily understood. The length of the hydrophobic chain has a direct relationship to the solubility of the molecule. A large carbon chain results in a reduction in the polarity of the molecule. This, in turn, increases the tendency of the molecule to distance itself from an aqueous environment, driving aggregation and as a consequence causing a reduction in the cmc. The presence of counter-ions in this case contributes to the reduction in repulsion between charged hydrophilic heads on the surface of the micelle and reduces the cmc. As a very rough guide organic counter-ions reduce the cmc by a greater amount than inorganic counter-ions, and divalent ions by a greater amount than monovalent ions. The presence of salt lowers the cmc of ionic micelles for reasons analogous to those mentioned in reference to counter-ions. Increases in temperature lead to an increase in the solubility of the molecules, as we will see on the Krafft point. Later we will see that the presence of less polar solvents leads to fundamental changes in the solvation and ultimately to the form of the aggregates (e.g., with benzene as a solvent we would expect inverse micelles with the hydrophobic tails turned toward the outside as the lyophilic parts and the hydrophilic heads congregated on the inside of the micelle).

A third parameter of particular technological interest is the Krafft point, which relates to the behavior of surfactants as a function of temperature as the cmc relates to the concentration. The Krafft phenomenon has to do with the sudden change in solubility of surfactants at a particular temperature that is called the *Krafft point*. At temperatures below the Krafft point, the surfactant is minimally soluble in water. As soon as the temperature passes the Krafft point, its solubility increases dramatically (some tens or hundreds of times in a space of a few degrees). The reason this happens is that in cases where the dissolved concentration of a surfactant molecule becomes close to the cmc, the molecules in question when warmed will switch directly from the nondissolved form to the formation of micelles with the attainment of the Krafft point. Accepting that the solubility of micelles is independent of temperature, with this mechanism the direct transformation from insoluble monomers to dissolved micelles is achieved upon attaining a particular temperature.

5.4   Deviations from the spherical micelle

Up to now we have been discussing the simplest version of the micelle. The basic assumptions made in the previous section were that[*]

- The individual surfactant molecules are comprised of a hydrophilic head and a comparatively large hydrophobic tail.

---

[*] For a thorough presentation of this very interesting topic, see, i.e., Evans and Wenneström (1994); Holmberg et al. (2003); Cosgrove (2005).

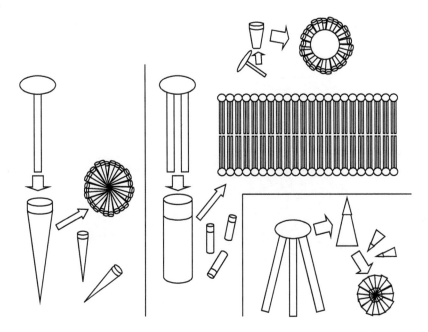

Figure 5.2 Approximate geometric shapes of surfactants. Note how the shape (and the charge) of the surfactant determines the shape of its aggregation.

- Because of the above, the surfactant molecule can be considered to resemble a cone, with the base as the head and the apex as the tip of the tail.
- The tails were assumed to be rigid.
- The molecules in an aqueous environment will aggregate so that the apices of the cones meet in the center of a sphere that is delimited by the hydrophilic bases of the cones.
- The aggregation number equates to the number of individual cones that can aggregate to form a sphere.

In reality, there are many cases in which one or more of the above assumptions is incorrect. In particular,

- Surfactant aggregation is sensitive to the presence of salts, especially in the case of charged surfactants.
- Many surfactants have more than one hydrophobic tail.
- The breadth of the tail is not necessarily smaller than that of the head.
- The shape of the molecule may not resemble a cone, but be more satisfactorily compared to other geometric shapes.
- The shape and aggregation pattern is sensitive to temperature.

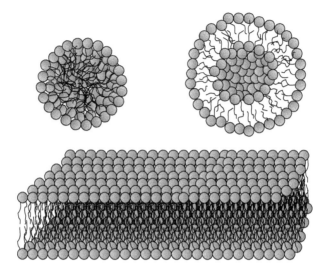

Figure 5.3 Schematic depiction of self-organized structures in cross-section. Above: micelle (left), liposome (right). Below: lamella. Note how the two last structures consist of successive layers of surfactants with those of the liposome showing curvature and those of the lamella being straight.

- The shapes in question vary according to the relative dimensions of the head and the tail, the shape of the tail, and the area of the head. Usually, the resulting shapes are cones, truncated cones, cylinders, or hollow cylindrical sheets.
- Just as the joining of cones leads to the formation of spheres, in the same way the joining together of nonconical elements leads to the formation of other three-dimensional geometric structures (see Figure 5.3).

5.5   The thermodynamics of self-assembly

By self-assembly we mean the spontaneous aggregation and shaping of disordered material into organized entities. The word "spontaneous" means that this process happens without the intervention of external factors such as catalysts or any form of supplied work. The simplest approach to the spontaneous creation of a structure is the study of the creation of a micelle.

The change in energy during the creation of micelles ΔG_{mic} can be attributed in general terms to the sum of free energy changes relating to (1) the contact of the hydrophobic tails of the surfactant with the water at the surface of the micelle (ΔG_{cont}), (2) the restriction of the hydrophobic tails in the center of the micelle (ΔG_{agg}), (3) the transfer of the hydrophobic tails of the surfactant from the continuous phase to the hydrophobic interior of the micelle (ΔG_{tail}), and (4) the electrostatic and stereochemical interactions between the hydrophilic/charged heads at the surface of the micelle (ΔG_{head}):

$$\Delta G_{mic} = \Delta G_{cont} + \Delta G_{agg} + \Delta G_{tail} + \Delta G_{head} \qquad (5.1)$$

Now consider an accumulation of surfactants, during which isolated molecules aggregate to form micelles. We can describe the formation of micelles as a hypothetical condensation reaction of "monomeric" surfactant molecules T to a "polymeric" micelle:

$$T + T \rightleftharpoons T_2 + T \rightleftharpoons T_3 + T... \rightleftharpoons T_n + T \rightleftharpoons... \qquad (5.2)$$

Because it is difficult to foresee a common equilibrium constant for every individual reaction, a simplified (and only partly effective but adequate for teaching purposes) approach is to consider that the value K_m can describe all the individual condensations of "monomeric" surfactants T onto the "polymeric" aggregate T_n that will develop into a micelle:

$$nT \rightleftharpoons T_n \qquad (5.3)$$

If we want to further consider micelle formation as the separation of a new phase consisting of n molecules from a population ($n + m$) of initially dissolved surfactant molecules, then the following results:

$$(n + m)T \rightleftharpoons mT + T_n \qquad (5.4)$$

In both cases, a final equilibrium is established between free surfactant concentration [T] and "polymeric" surfactant concentration:

$$K_m = \frac{[T_n][T]^m}{[T]^{n+m}} = \frac{[T_n]}{[T]^n} \qquad (5.5)$$

The standard energy of micelle formation ΔG_m^0 is given by the relationship

$$\Delta G_m^0 = -RT \ln K_m \qquad (5.6)$$

$$\Delta G_m^0 = -RT \ln[T_n] + nRT \ln[T] \qquad (5.7)$$

and dividing by the number of molecules per micelle n we obtain the standard energy of micelle formation for 1 molecule of surfactant:

$$\frac{\Delta G_m^o}{n} = -\frac{RT}{n}\ln[T_n] + RT\ln[T] \qquad (5.8)$$

For large aggregations of molecules per micelle (i.e., more than 100 molecules), the first term on the right-hand side of Equation (5.7) can be omitted. Assuming that micelles will form when the concentration of monomers becomes equal to the cmc (Section 5.3), the standard energy of micelle formation for 1 mol of surfactant $\Delta G_{M,m}^0$ approximately equals

$$\Delta G_{M,m}^0 \approx RT\ln[T] \qquad (5.9)$$

$$[T] = cmc \qquad (5.10)$$

$$\Delta G_{M,m}^0 \approx RT\ln cmc \qquad (5.11)$$

If the surfactants are charged, then we must take into account the surface charge, which relates to the charge of the surfactant x and the total concentration [A] of counter-ions A with charge y that exist in the solution. Of these, a quantity with concentration p will accumulate on the surface of the micelle in order to balance its charge. The chemical equilibrium then should take into consideration the dissociation of the surfactant.

The standard entropy of micelle formation is given by the relationship

$$\Delta S^0 = -\frac{d(\Delta G^0)}{dT} \qquad (5.12)$$

$$\Delta S^0 = -RT\frac{d(\ln cmc)}{dT} - R(\ln cmc) \qquad (5.13)$$

and the standard enthalpy of micelle formation for uncharged surfactants is equal to

$$\Delta H^0 = \Delta G^0 + T\Delta S^0 \qquad (5.14)$$

$$\Delta H^0 = -RT^2\frac{d(\ln cmc)}{dT} \qquad (5.15)$$

The next question that must be posed concerns the form of the condensate T_n. The main parameter that determines the form of the self-assembled structure is the geometric shape that best approximates the shape of the molecule. This simplified approximation cannot take into account electrostatic or hydrophobic interactions between the molecules, but despite that, it gives very interesting insight into the way in which nonspherical structures are formed. Let us consider a collection of similar molecules, the shape of which can be approximated to a cone with the base for the head and the apex for the tail (case A) and another equivalent collection with a shape that approximates to a cylinder, for example, a surfactant with two tails (case B).

It is obvious that in case A, the surfactant molecules will form a spherical aggregate in accordance with classical micelle theory. The surfactants in case B cannot aggregate into a sphere, but order themselves in parallel and form a plane with a thickness equal to the length of the cylinder.

We mentioned previously that the radius of the micelle r approximates to the length of the tail l of the surfactant. If we consider N the aggregation number (molecules/micelle), a the part or the area of the micelle that corresponds to 1 molecule, and v the volume that corresponds to the geometric shape that describes the molecule (e.g., a cone), then for the ratio of the volume of the micelle V to the volume of 1 molecule v we have:

$$N = \frac{V}{v} = \frac{4}{3}\frac{\pi l^3}{v} \qquad (5.16)$$

while for the ratio of the surface of the micelle A to the surface of 1 molecule a, we can write

$$N = \frac{A}{a} = 4\frac{\pi l^2}{\alpha} \qquad (5.17)$$

Equating Equations (5.16) and (5.17) gives us:

$$\frac{a}{v} = \frac{3}{l} \Rightarrow \frac{v}{la} = \frac{1}{3} \qquad (5.18)$$

which is the mathematical description of a cone.

For a spherical micelle, the following applies:

$$\frac{v}{la} \le \frac{1}{3} \qquad (5.19)$$

The combination of many cones like surfactants creates a spherical micelle (Figure 5.3). The same reasoning can be used for the determination of the critical packing parameter P_{cr}:

$$P_{cr} = \frac{v}{la} \qquad (5.20)$$

A volume of work by Tanford, Israelachvill and others correlated the shape of the molecule to the resulting structure. Surfactants with two hydrophobic tails (e.g., lecithins) tend to form structures with less spontaneous curvature than surfactants with one hydrophobic tail (e.g., dodecyl sulfonic salts), and therefore tend to form sheets rather than micelles. As shown in Figure 5.3, the geometry of a surfactant molecule can give a good idea of the structures that it is likely to form.

5.6   *Structures resulting from self-assembly*

In previous sections we referred to the self-assembly of surfactants into structures, the most characteristic of which is the micelle. We also mentioned deviations from the spherical form of micelles, with a general reference to alternative structures. Here we embark on a more extensive description of these structures, as they are of exceptional technological and biological interest. Previously we mentioned that the empirical criterion that correlates the structure of the molecule with the structure of the aggregate that will be created is the approximation of the form of the isolated molecule to some three-dimensional geometric shape. Molecules with the shape of cones, cylinders, inverted truncated cones, and inverted cones lead to the formation of micelles, hollow micelles or membranes, inverse hollow micelles, and inverse micelles, respectively. The sequence of these structures coincides with a gradual change in the spontaneous curvature from positive values to zero and then to negative values.

Likewise, we have made extensive reference to the fact that the form of self-assembled structures depends on the concentration of the surfactant, the temperature and the existence of other solutes. These relationships are understandable from the phase diagrams. The surfactant molecules organize themselves into macroscopic structural arrangements throughout the solution in which they are found. This can occur up to very high concentrations, after which phase separation occurs. Saturation is not unusual in surfactant solutions. In the case of surfactants with a single hydrophobic tail, this is determined to a large degree by the length of the tail. Surfactants with a small number of carbon atoms in the tail tend to maintain the spherical micelle structure at high concentrations even up to the point of supersaturation. As the number of atoms in the carbon chain of the tail increases, the surfactant molecules tend to group together in long tube-like aggregates with circular cross-sections, effectively forming elongated micelles.

The tendency to form such structures (rod-shaped micelles), which can in essence be fibers tens or hundreds of nanometers in length that extend throughout the solution, becomes stronger as the number of carbon atoms in the hydrophobic tail increases. The formation of such rod-like micelles increases the viscosity of the solution. They effectively behave as

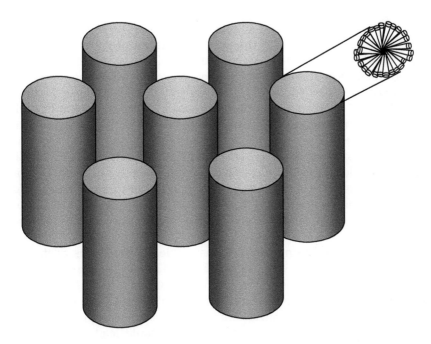

Figure 5.4 Schematic depiction of the hexagonal phase. Every cylinder consists of a repeating series of circular arrangements of surfactants.

polymers, and the behavior of such solutions has been interpreted with reference to polymer solutions. The most basic deviations from the behavior of polymer solutions are the very high dependency on the concentration of the surfactant, the temperature, and the particularly interesting property of forming branches which can lead to complicated structure. Many mechanisms have been proposed for this interesting property.

These elongated micelles differ in elasticity and flexibility, so that they can be defined as rigid, semi-flexible, and flexible rods. In the case of ionic surfactants, the presence of electrolytes plays a major role in their rigidity. As is the case for polymer solutions, we can determine the *semi-dilute limit concentration*, before which the rod-like micelles move freely in the liquid without significant interactions between them. In polymer physics in such cases, the hydrodynamic radius of an isolated polymer and the corresponding size distribution formula (usually a bell-shaped Gaussian-like curve) can be defined. Such structures present an interesting and sometimes complicated rheological properties. Here we should remember that the basis for the rheology of colloidal dispersions is Einstein's well-known equation.

$$\eta = \eta_o \, (1 + 2.5 \, \varphi) \tag{5.21}$$

where η is the viscosity of the system, η_0 is the viscosity of the solvent, and φ is the volume fraction of the dispersed substance. Of course, this simple theoretical equation fails to describe the rheology of complex or concentrated dispersions. Significant complications occur when the probability that two dispersed particles ("particles" refers here to the rod-shaped micelles) will approach each other increases with the concentration of the surfactant (and therefore of the micelles). We can define the semi-dilute limit at surfactant volume fraction φ^* from which point and above we cannot assume that there are no mutual interactions between the dissolved micelles. The very large deviations from the simple model of concentration–viscosity show that the dominant factor in the determination of the viscosity is not the polymers/rod-shaped micelles themselves but rather the mutual interaction between two such entities as they approach one another.

The shape and form of the self-assembled structures is related not only to the shape of the surfactant, but also to its temperature.

To walk through the various forms of surfactant self-assembly, and also to highlight their generic relation to concentration, we can imagine an abstract paradigm of a surfactant added so as to form spherical or rod-like micelles. Further addition of surfactant could lead, under certain conditions, to the formation of a cubic phase, which is an isotropic bi-continuous state. *Bi-continuous states* are those states in which two continuous phases coexist in the same system. As a typical example, we can take a sponge in which all the pores are interconnected. Continuous phases in this case are the pores and the solid material that surrounds them. Alterations in the structure of the surfactant leads to the formation of a *lamella*, which is another anisotropic structure comprised of molecules of surfactant with no curvature. It is the generally preferred self-assembled structure for surfactants for which the geometry of the molecules approximates to a cylinder (e.g., molecules with two hydrophobic groups), but noncylindrical molecules in high concentrations can form such structures. In this progression (micelles → hexagonal phase → cubic phase → lamella), the spontaneous curvature is transformed by high concentrations to zero (the surface of a micelle is very curved, of the cubic phase a little curved, and of the lamella not at all)—always with the hydrophilic groups outward. Still further changes in the structure (i.e., hydrophobicity) of the surfactant causes a phase inversion in which the surfactant is present in greater quantities than the water and it can be considered the solvent with the water dissolved in it especially in high surfactant concentrations. The water is enclosed in pockets inside the surfactant structure, where at the interfaces the surfactant will have the hydrophilic groups toward the inside (toward the solvated water). In this way, the spontaneous curvature becomes negative, with the formation of the isotropic inverse cubic phase (which is not bi-continuous, because the water pockets

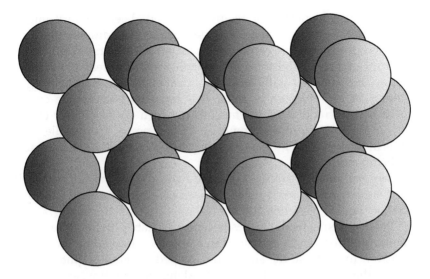

Figure 5.5 Schematic representation of the arrangement of hydrophobic heads in the cubic phase. The structure can be bicontinuous, that is, the cores of the spheres can communicate with one another.

are not interconnected (Figure 5.5) and the even more curved anisotropic inverse hexagonal phase.

A more analytical discussion of the individual phases will be useful, as they are of huge significance in technology and biology.

5.6.1 Spherical micelles

We do not discuss these extensively, as they were examined in previous sections. For review purposes, it must be emphasized that these entities are formed from molecules whose shape approximates to a cone (one hydrophobic tail) in generally low concentrations. In micelles, the specific surface of the molecule must be fairly large, without the radius of the micelle r exceeding the length of the hydrophobic tail l. To satisfy the need for a large value of a (area of the head), the formation of micelles is facilitated by charged heads, in order that the repulsions between heads can increase the specific surface area that one head in the micelle occupies. The general conditions for such a micelle formation are $C \Leftarrow l$, $v/(al) \leq 1/3$.

5.6.2 Cylindrical micelles

These are formed by molecules that do not have a sufficiently large head area to form spherical micelles. The geometry of molecules with 0.33–$0.5 \leq v/(al) < 1.0$ approximates to a truncated cone and such molecules generally form rod-shaped micelles or relatively curved sheets. Here it must be stressed that the same molecules under different conditions of salt concentration or temperature can change geometry, for example, from a cone to a truncated cone, and thus self-assemble into different structures. In this way, SDS, which usually forms spherical micelles, can form elongated cylinders in the presence of salts. This happens because the counterions of the salt reduce the interactions between the negatively charged heads of the surfactant that comprise the surface of the micelle, which results in the heads packing more tightly together and the specific surface area being reduced.

Here it should be stressed that the hydrophobic interior of micelles and of similar structures is not comprised of straight ordered chains, but rather of "liquid" of carbon chains.

The size of spherical micelles is generally stable, because the number of molecules that can be combined in a micelle is limited by their stereochemistry. In contrast, the size of rod-shaped micelles is quite variable, because the extension of the cylinder along the lengthwise axis is not subjected to the severe restrictions to which the radial extension is subjected. In simple words, while an individual molecule would not be able to "squeeze in" and increase the cross-sectional area of the cylinder, it can always attach to one of the ends, thereby lengthening the whole structure. The mean aggregation number (which for a constant cross-sectional area is proportional to the length) is dependent on the concentration C and the specific area a:

$$\langle N \rangle = 2\sqrt{C \exp a} \qquad (5.22)$$

This means that the mean aggregation number depends on all of the parameters that influence a, such as salt concentration, temperature, pH, etc. (Figure 5.4).

5.6.3 Lamellae: Membranes

Lamellar structures such as membranes are formed when the geometry of the surfactant molecules is described by the relationship $v/(al) \approx 1$, that is, when we have an open truncated cone or, better still, a cylinder. This can happen either because the specific surface area has fallen to a low value due to the effect of ions or temperature changes, or because the lyophobic tail is too bulky to be considered the apex of a cone (e.g., in the case of two lyophobic tails). The lyophobic parts are ordered with one opposite the other in order to form a sheet two molecules thick with the lyophilic parts oriented toward the outside. Typical examples of molecules that

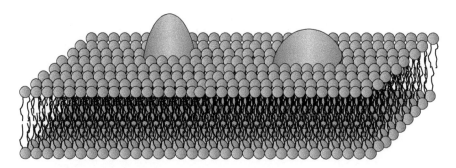

Figure 5.6 Three-dimensional section of a membrane produced by a lamella of the two-tailed surfactants with interspread proteins.

approximate a cylindrical form and self-order into sheets are many of the phospholipids. These substances form membranes even at low concentrations without passing through the micelle stage. Furthermore, thermodynamic factors impose a depression of the cmc in the case of molecules with two tails. One can conceive that, with the addition of the first few phospholipid molecules to water, membranes spontaneously form when the concentration of free phospholipids (outside the membranes) is very low. This peculiarity of phospholipids constitutes one of the most important properties of biological material. Many cell membranes are essentially self-assembled phospholipid membranes with proteins that are embedded in order to perform particular functions (Figure 5.6). The properties of these membranes include selective permeability and great elasticity.

5.6.4   Hollow micelles

In concentrated solutions of surfactants, the membranes are clustered in densely stacked layers. These layers are called lamellae and form parallel plates with interposing layers of water. Liquid crystals are associated with such forms. In fact, applications of liquid crystals are one of the main topics of research today.

The flat surface of a membrane that is formed from cylindrical structural elements shows no curvature. The problem here is that such structures have aggregation numbers that tend to infinity (they have no end), an undesirable fact from the entropic point of view. Flat surfaces logically will try to close off their open edges in order to obtain a finite form and aggregation number. This can occur if the components at the edges can aggregate and the structures fold into closed spherical membrane (vesicles), essentially hollow micelles. In order for this to take place, the molecules of the external layer must deviate from a completely cylindrical shape, that is, their value of $v/(al)$ must fall between 0.5 and 1.0 so that the surface will curve more than the inside. Theoretical calculations have shown that there is a critical radius R_c from which point and above the structure can withstand stereochemical stresses due to the shape distortion without the spherical lamella being burdened energetically:

$$R_c = \frac{l}{\left(1 - \dfrac{v}{al}\right)} \qquad (5.23)$$

It is speculated by some that spontaneous structures such as hollow micelles and lamellae present on the primitive Earth may have been the ancestors of living cells. These structures, which have practically the same structure as cell membranes, could have formed spontaneously from isolated surfactants. Thus, it is speculated that on the ancient Earth, closed spherical lamellae (hollow micelles, vesicles/liposomes) somehow trapped some organic or silicate genetic material with the capacity to self-replicate, essentially becoming the first prokaryotic cells.

5.6.5   Inverse structures

Many of the structures referred to previously have two states. One of these is found when water is the continuous phase of the system and the surfactant can be considered "dissolved" in it. The other is that which is obtained when the surfactant is the main component of the system. This generally occurs when the hydrophobic tail is bulkier than the hydrophilic head ($v/al > 1$). In this case, the water can be considered to be dissolved in the continuous surfactant phase. Theoretically, one can encounter all of the "basic" self-assembled structures (there are many others which are found more rarely but are still of some importance) if one starts with pure water and adds surfactant gradually until the concentration of the water tends toward zero. The result can be presented schematically with the following sequence, often quoted as the Fontell diagram:

> Micelles < Cubic phase < Hexagonal phase < Spherical lamella < Lamella < Inverse spherical lamella < Inverse hexagonal phase < Inverse cubic phase < Inverse micelles

The spontaneous curvature is zero in the lamella. Cubic phases can be found between most transitions from one phase to the other.

The Fontell diagram, which is comparable to the progression previously described in the paragraph on the stereochemistry of molecules

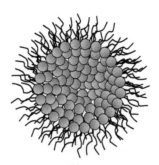

Figure 5.7 Surfactant with two hydrophobic tails whose shape resembles a cone with the hydrophilic head at the apex. (see Figure 5.3). The obvious structure into which they will self-assemble is a micelle with the tails turned outward. Can this provide evidence for the HLB of the molecule?

and the structure of self-assembled entities, orders the structures by spontaneous curvature with the noncurved sheets found in the center. Left and right are mirror images of the same structures but with their phases inverted. The succession of structures from micelle to lamella or from reverse micelle to lamella corn, in cases, results from increase in the concentration of the surfactant follows the spontaneous curvature from highly positive to highly negative values. Hence, if further surfactant is added to a system that contains micelles, strong repulsive forces between the micelles drive them as far away from each other as possible. If the micelles then reform themselves into rod-shaped micelles that are arranged in the hexagonal phase, their surfaces will distance themselves as much as possible from each other. Similar successive phases can lead to the formation of the cubic phase then to lamellae. It is possible for a surfactant to not pass through all these phases while its concentration is increasing, because, in the discussion above, the role of temperature was not accounted for. Here it must be remembered that the above is a generic paradigm and a somewhat abstract description of the phase transitions of an "ideal" surfactant, which of course does not exist. For a realistic prediction of the behavior of a water–surfactant system, the phase diagram of the substance in question must be consulted.

Inverse micelles tend to be smaller than normal micelles, at least in the case of ionic surfactants, which as a general rule have smaller hydrophilic sections. The usually small amount of water present in such systems is found solvated on the inside of the micelles while the continuous phase consists of the aggregation of the hydrophobic tails, which solvate each other. In this way, a system of inverse micelles can be considered an emulsion of water dispersed in oil. The occasional change from micelles through the other phases to inverse micelles can be considered a phase inversion (water in oil → oil in water) and can be observed with techniques that measure properties that alter from one phase to the other, such as conductivity, because the continuous aqueous phase can conduct electricity and the continuous noncharged, nonpolar organic phase of inverse micelles cannot.

5.7   Phase diagrams

In previous sections, sporadic references were made to phase diagrams. These are perhaps known to the reader from basic thermodynamic phase diagrams that correlate the pressure of a material with its volume, defining its physical state in relation to these variables. In the science of polymers, surfactants, and heterogeneous systems generally, phase diagrams serve a similar purpose: On the basis of two or more variables, a map is created with which one can predict the physical state of a system. The variables in this case are the concentrations of the individual components and the temperature. For a simple two-component system (e.g., water and surfactant), the phase diagram has two axes: the concentration (essentially the proportion of the two components) and the temperature. Every point on the diagram corresponds to a solution of a particular concentration and at a particular temperature. For three-component systems (e.g., water, a surfactant, and another solvent or surfactant), the phase diagram will be a triangular Gibbs diagram. In these diagrams, every corner of an equilateral triangle corresponds to one component and every side to the individual concentrations in the system. Usually, these phase diagrams are given for one temperature close to room temperature, for example, 25°C. If the temperature is to be included in the diagram, then there will be an axis perpendicular to the triangle so that the diagram becomes a prism. Due to the intrinsic difficulties of orientation of the phases and the construction of the corresponding diagram, few prismatic diagrams are available, most of them covering a limited range of temperatures and compositions.

5.8   Self-assembly of macromolecules:
The example of proteins

The issue of the solubilization of proteins in aqueous solutions and of their conformations and interactions therein is an extremely complicated physicochemical topic, the scope of which is far beyond the aim of this book. What follows is a simplified approach, aimed only at introducing the basic themes of protein solution chemistry, and building a link between the basic notions of physical chemistry introduced earlier in this book and complex physicochemical entities such as proteins.

What happens when, instead of hundreds of free molecules, we have a very large number of individual molecules joined together in a polymer chain? This situation initially discussed in Chapter 3 directly concerns polysaccharides (polymers of sugars) and proteins (polymers of amino acids). In the case of polysaccharides, the abundance of particularly polar hydroxyl groups, −OH, usually ensures the solubilization of macromolecules. A marked (and common) exception occurs when the interactions due to the formation of hydrogen bonds between neighboring polysaccharides are very powerful. In that case, such hydrogen bonds between adjacent polymers will not break easily, and neighboring polysaccharide chains will not readily dissociate in order to hydrate (that is, to establish new hydrogen bonds with water). Such phenomena render the solubilization of polysaccharides such as cellulose and chitin difficult.

In proteins, the issue is more complicated because both polar and nonpolar amino acids can be found together in the protein macromolecule. Proteins with high polar amino acid content tend to have elongated, unfolded forms in order to maximize the interaction of these amino acids with the water. On the contrary, proteins with high nonpolar amino acid content tend to adopt almost spherical configurations in order to minimize interaction with the water. This tendency of the nonpolar amino acids to move away from water leads them into the interior of the protein structure, while the polar amino acids remain at the outside. This ideally results in a folded structure with a hydrophobic core and a hydrophilic surface.

The above suggests macromolecules with discrete long segments comprised exclusively of either hydrophilic or hydrophobic amino acids. In reality, extended sequences of highly hydrophilic or hydrophobic amino acids are rare. In a typical protein, hydrophobic groups are expected to be found between hydrophilic groups on the external surface of the tertiary structure, while hydrophilic groups can be found on the inside of a folded protein.

Often, the hydrophobic groups are so dense even on the external surface of the tertiary structure that they interact with the hydrophobic groups of *other* proteins in order to avoid the aqueous environment. Such supermolecular structures in nature usually consist of from two up to several tens of proteins. These *quaternary structures* can strengthen themselves by forming disulfide bonds between free cysteine residues on neighboring proteins and are understandably common in proteins rich in cysteine, such as flour gluten. The mechanical durability of flour is related to its cysteine content, because the many covalent −S−S− disulfide bonds join the proteins strongly to one another, imparting a "hard" texture to different bread products.

5.8.1   Why are all proteins not compact spheres with their few nonpolar amino acids on the inside?

Let us consider the protein as a material body with a surface occupied by both hydrophilic and hydrophobic amino acids. Folding of that structure would be required in order for the hydrophobic parts to be removed from the outer part. The thermodynamic incentive for that folding is a reduction in free surface area G_{surf}. Perfect folding of the structure into a sphere does not occur because there are other energy factors that oppose this. Chief among these is the energy $T\Delta S_{conf}$ that relates to the conformational entropy ΔS_{conf}: Perfect folding and compacting of the sections of the peptides on the inside of the structure would seriously limit the internal mobility of the protein chain. This entails a reduction in the degree of freedom in the macromolecule, and consequently an undesirable reduction in entropy. In this simple scenario where a number of other factors are omitted for the sake of simplicity, the degree of folding of a protein is ultimately a compromise between

1. The reduction in free surface energy ΔG_{surf} due to the distancing of nonpolar amino acids from the water, which facilitates the folding
2. The increasing of the conformational entropy ΔS_{conf} due to the free movement of the sections of the protein chain inside the tertiary and quaternary structure[*]

5.8.2   How do proteins behave in solution?

The forces that keep the protein molecule dissolved (suspended) in an aqueous environment are due to the hydrogen, ionic, and other bonds that form between ionized amino acids and other polar moieties on the external surface of the protein and the water. According to this rationale, the forces acting toward the solubilization of the protein are stronger when the protein is folded, so as to maximize the concentration of the polar amino acids on its interface. Later on we will see that unfolding of the protein and exposure of the hydrophobic parts can consequently reduce its solubility.

Amino acids with acidic substituents ionize at pH values higher than their pK values (Section 3.10), giving negatively charged ions. Those with

---

[*] In reality, there are more thermodynamic factors that compete for and against the folding, and these relate to the free energy contribution from hydrogen bonds, electrostatic forces, van der Waals forces, and London forces, among others. Some of them, such as intra-macromolecular hydrogen bonds, can be the dominant forces involved in folding. Despite this, the simplified approach in the text describing the competition between free surface energy and conformational entropy at a given temperature is satisfactory for teaching purposes. For more details, see, i.e., Dickinson (1994).

basic substituents ionize at pH values lower than their pK, giving positively charged ions. As we can imagine, reducing the pH of a protein solution leads to a gradual reduction of negative charges as the acidic groups approach their pK values and change from a charged to an uncharged state while the basic groups begin to ionize. As mentioned in the discussion on polymers, there will be a particular pH value at which the cumulative positive charges of a protein are equal to the cumulative negative charges. The value of this *isoelectric point* pI is different for every protein and depends on the amino acids of which it is comprised.

Proteins become suspended in aqueous solutions primarily due to a range of interactions between the water molecules and the polar amino acids, in which charges could play some significant role. Furthermore, the proteins do not approach one another but remain dispersed in the aqueous medium due to the electrostatic repulsions arising from their charged groups. At the isoelectric point, the proteins have no net charge. As discussed below, the negation of this net charge causes a phase separation between water and protein, resulting in precipitation of the protein. In addition, because the surface charge of the protein has been removed, the electrostatic repulsions between the proteins that were due to the similar charges are also eliminated, and the proteins flocculate.

Summarizing, the removal of ionization of a protein causes flocculation because, among many other factors, some of the following occur:

- The forces between polar amino acids on the protein surface are, in the absence of electrostatic repulsions, significant; hence, there is a clear attraction between uncharged proteins.
- The formation of hydrogen bonds preferably between water molecules leads to an increase in the free energy between the water and protein surfaces, resulting in the flocculation of the protein in order to reduce these unfavorable interactions (reduction in protein–water interface tension). The attentive reader will easily understand the similarity of this mechanism with the folding of the hydrophobic groups inside the protein.

In both cases, the energetic gain derives from the formation of bonds between the water molecules that would otherwise find themselves confronted with the uncharged surface of the protein.

Precipitation and flocculation of proteins by means of bringing the pH to the isoelectric point are widely used techniques in food technology and cooking. The preparation of a number of dairy products relies on the flocculation of proteins as the pH of the milk falls from its normal value of ~7 to the isoelectric point of milk caseins, a complex group of milk proteins. The acidification takes place via a number of routes, such as the induction of fermentation reactions, which include microbial production of lactic acid or the incorporation of δ-lactone of gluconic acid (glucono-δ-lactone). The hydrolysis of this substance causes a gradual reduction in pH. The casein flocculates and the products, casein condensate, colloidal calcium phosphate, and fat droplets onto which casein could be adsorbed are further processed as required for the preparation of a range of dairy products. In cooking, the preparation of egg and lemon sauce requires the continual control of pH in order for the pH not to fall below the isoelectric point of the albumin and curdle the sauce (flocculate the egg proteins).

5.8.3   *A protein folding on its own: The Levinthal paradox*

Let us consider the reversible reaction between the unfolded and the folded state of a protein:

$$\text{Unfolded protein} \rightleftharpoons \text{Folded protein}$$

As with the simpler chemical reactions studied in the previous chapters, an energy term is associated with it. As this is a reaction involving not two or three atoms, but the organized movement of hundreds of atoms, the energetic content of the reaction cannot be readily described in terms of enthalpy alone ("bonds"), but must also include entropic terms (organized movement of molecular ensembles such as a peptide chain). It is thus appropriate to consider protein folding in terms of free energy $G = H - TS$ rather than of enthalpy H alone. For the sake of simplicity, we can omit the interactions with the solvent and the effect of free surface energy that we encountered earlier. In the present case, protein folding will occur only due to direct intramolecular interactions between amino acids (hydrophobic amino acids react under the influence of dispersion forces; hydrophilic amino acids under a wider range of polar interactions). As a first approach, for the folding to occur spontaneously the free energy associated with the transition from the totally folded to the natural (partially unfolded) state, ΔG_{conf} must be negative. As folding will reduce the mobility, and hence the entropy, it must be energetically compensated for by the formation of bonds, that is, to decrease the system enthalpy. Considering that, at least initially, folding and bond formation reduce H to a larger extent than they reduce TS, it is reasonable to assume that a completely unfolded protein will undergo a series of successive foldings: In each of the foldings, H is going to decrease due to the formation of new bonds, while S is going to decrease due to the restriction in the chain mobility brought about by this folding. As the prime sites for bond formation become saturated and as the chain mobility becomes more and more restricted, the difference between $\Delta H_{conf} - T\Delta S_{conf} = \Delta G_{conf}$ will approach zero and the folding will stop. The final, ideal structure is a trade-off between the formation of many bonds (minimization of enthalpy) and the retention of the maximum possible freedom of movement (maximization of entropy). Theoretically, every protein left in a completely unfolded state will revert to the conformation for which $\Delta G_{conf} = 0$. However, in the late 1960s, Levinthal published his famous paradox, calculating that the steps to be taken for a protein to fold from a completely unfolded state would require an immense amount of time. However, in reality, proteins tend to fold in far less than a second. The prime reason for this is that proteins do not fold on a residue-by-residue basis; rather, they first fold on a local scale, and such structures then fold within themselves, the resulting structures folding within themselves and so on, dramatically decreasing the time needed for folding. Another very important consideration is that, in some cases, proteins do not appear to reach the point where $\Delta G_{conf} = 0$. It appears that thermodynamically unstable conformations can be stabilized *kinetically*; that is, they reach a point that is stable for any practical application, and from which any changes toward further reduction in free energy would take too long to be perceived by any observer.

5.8.4 *What happens when proteins are heated?*

The total free energy gain ΔG_{conf} for the folding of a protein (i.e., for the transition from an unfolded to a folded conformation) is small, on the order of a few tens of kilocalories per mole (kcal mol^{-1}): Small variations in temperature increase the motility and consequently the conformational entropy, which, as we have seen, facilitates the unraveling of the quaternary and tertiary protein structure. The secondary structure, stabilized by strong interactions (hydrogen bonds, stereochemical interactions) is less sensitive to heating.

Hence, when energy is provided to a protein in the form of heat, mechanical energy in the form of vigorous shaking, or in other ways, the fine balance of forces that maintains the tertiary and quaternary structure can be disrupted. An increase in temperature will lead to an increase in the entropic factor that corresponds to the conformational entropy $T\Delta S_{conf}$. This factor, as we have seen, favors the unfolding of the chain, so the provision of heat causes the chain to unfold. This loss of tertiary and quaternary structure is called *denaturation*.

Theoretically, if after cooling the protein returns to its original temperature (i.e., the balance between the surface free energy plus any enthalpy due to intramolecular interactions and the energy from the conformational entropy is restored), the protein will return to its initial folded state. Indeed, this has been observed on many occasions, especially for very dilute protein solutions. In practice, however, if milk is scalded or liver hardens during cooking (changes in texture caused by proteins denaturing), as much as they cool, they will not return to their original state.

In food technology we know that if we cut and blanch a vegetable, the browning-inducing enzymes are denatured. If we let this vegetable cool, it will not readily brown: This is related to the deactivation of the enzymes. Our overall experience from cooking and food processing tells us that denaturation can be irreversible. Although a limited number of enzymes have been shown to recover part of their activity following denaturation, it seems that the destruction of particular sections of secondary structure can be irreversible.

Denaturation is often followed by the flocculation of proteins. Flocculation can render any return to the native state even more difficult. Flocculation leads either to the precipitation of proteins, such as in the heating of caseins and lactoglobulins (scalding of milk), or to the creation of gels (upon denaturing, the protein spreads out in the space available and creates a mechanically strong network), such as in the gelling of gelatin or of albumin during the boiling of an egg.

It is generally accepted that the denaturation of proteins with a spherical structure (enzymes, globulins, albumins) progresses from the initial conformation to the denatured, open structure via an intermediate stage, called a *molten globule*. This can be a reversible stage, during which there is partial exposure of the hydrophobic groups, but in some cases the initial structure can at this stage be regained with the restoration of the initial energetic equilibrium, (i.e., by cooling).

5.8.5   What is the effect of a solvent on a protein?

It is usual for most typical proteins to be considered soluble in aqueous solvents. This is reasonable as the natural environment of a protein is in aqueous solution or at the interface of water with another medium, either in bodily fluids or tissues. Although in the laboratory the obvious first approach to dissolving a pure protein is to use buffered distilled or deionized water solutions of relatively low ionic strength, in an organism or a food proteins are usually dissolved in aqueous electrolyte solutions of high ionic strength.

An appropriate way to describe an ionic environment is with the activity of an electrolyte and with the whole ionic environment that is expressed by the ionic strength of the solution (see Section 3.11). In contrast with the concentration, the ionic strength of a solution takes into account the effects of the interaction between the ions. Small concentrations of electrolytes, as long as the pH does not change and bring about a decrease in the protein charge, generally enhance the solubility of proteins. Further increases in ionic strength normally lead eventually to the flocculation and precipitation of the proteins.

The ions compete for water molecules with the ionized parts of the protein surface, but also with the water molecules for the ionized parts of the protein. At low concentrations, the ions are bound by the surface charges of the protein and act as a charged sheath, thus improving the protein's solubility. At greater activity values, the ions are now so numerous that they out-compete the charged groups on the protein for water, eliminating protein–water hydrogen bonds and cutting off the protein from its aqueous environment. This phenomenon is called *salting out*.

5.8.6   What are the effects of a protein on its solvent?

Organisms use proteins to regulate the rheological behavior of various biological fluids. Man has also exploited the ability of proteins to alter the rheology of aqueous systems in several different fields. The most well-known protein that is used for such purposes is gelatin. Gelatin is a product of the thermal breakdown of collagen. The latter, a basic structural protein, irreversibly denatures and partially hydrolyzes on heating. The product, gelatin, traps the water with which it is in contact when it cools. The result is a significant number of immobilized water molecules. In the gel thus produced, the forces that keep the water molecules embedded in their places are stronger than those of gravitational attraction. We can therefore give shapes to gelatin gels that will not readily collapse under the effect of gravity.

Collagen, of which there are at least eleven different supermolecular structures, is an interesting example of the transition from the primary to the secondary and higher structures. The typical amino acid composition of collagen includes significant amounts of glycine, proline, and hydroxyproline, glycine making up close to one third of the total amino acids. As glycine has no side chains, it is used to facilitate the folding of the protein. The basic motif of the primary structure is –glycine–proline–X– or –glycine–X–hydroxyproline– (or similar structures), where X represents interchangeable polar or nonpolar amino acids so as to organize the folding and aggregation of the quaternary structures into higher super-structures, the *fibrils*. The resulting peptide chains are gigantic even for proteins, consisting of over 1,000 amino acids per chain. Most collagens appear to be comprised of three polypeptide chains that are twisted around one another in left-handed helices with three amino acids per revolution. This quaternary structure is stabilized with hydrogen bonds between the chains and also covalent bonds formed by the deamination of lysine. The latter increase with age and are largely responsible for the tougher texture of the meat from older animals.

Another series of proteins that can influence the rheological behavior of their solutions are the mucins, which are responsible for the mucous nature of saliva, gastric fluids, and tears. The mucins are glycoproteins that is proteins covalently bound with sugars, in which the peptide chain constitutes the backbone of the molecule. Covalently bonded oligosaccharides are found on the amino acids of the protein. These extend outward, forming a structure that, according to the currently predominant theories, resembles a hair brush with the body of the brush as the polypeptide and the bristles as the polysaccharides. In mucins, which are a very large family of glycoproteins, although the backbone of the molecule is a protein, the amino acids constitute only around 20% of the molecular mass (around 200 to 500 kDa out of 0.5 to 20 MDa). The remaining 80% consists of hydrocarbons, particularly oligosaccharides rougly made of five to twenty monomers, of which some can contain nitrogen.

The probably extended conformation of the protein backbone and the very large water binding capacity of the glycoside extremities of the molecule are responsible for the immobilization of large amounts of water and for the dramatic increase in the viscosity of the aqueous system forming the saliva or the gastric fluids. It is generally accepted that the mucin molecules cause flocculation to emulsions, causing the separation of phases into fat droplets and water, an occurrence that may be related to their creamy texture.

5.8.7 Protein denaturation: An overview

The above paragraphs referred to isolated cases of loss of structure of proteins due to thermal processing, change in pH, exposure to solutions of different ionic strength, or interaction with a hydrophobic surface. In order for a protein to denature in a first approach, the free energy for denaturation (ΔG_{denat}) must be negative. ΔG_{denat} will be equal to the difference between the free energy of the native state G_{nat} and that of the denatured state G_{den}.

The diligent reader will notice that thermal denaturation and the adsorption of a protein to an interface are thermodynamically and phenomenologically similar mechanisms that differ in the cause of the loss of balance between ΔG_{surf} and $T\Delta S_{conf}$: on heating, $T\Delta S_{conf}$ (entropic contribution from the conformation of the polymer) increases with the supply of heat, while in interface adsorption ΔG_{surf} (energetic change resulting from alterations in the interfacial interactions) is decreased by the better removal of hydrophobic groups from the aqueous environment. In both cases, the structure is lost because the total free energy is reduced.

Heating and surface adsorption then, together with change in pH or ionic strength, have a common root cause of protein denaturation, namely the disturbance of the fine balance of forces that drive the self-assembly of sections of the protein and the final manifestation of different structures. This suggests that the effect of heating, change in pH, surface adsorption, and other causes of denaturation are cumulative; joint application of high pressure and temperature can deactivate enzymes at temperatures lower than would be required with simple heating. The latter technology of successive combined treatments such as temperature and pressure (known as "hurdle technology") is a means to achieve a desired denaturation of proteins (cooking, deactivation of enzymes) without overheating the product, which could result in the loss of other desirable qualities and nutritional components.

Protein denaturation is of fundamental importance in food processing and cooking. Blanching is the thermal denaturation of enzymes such as phenyloxidases, which are responsible for the appearance of a brown color in cut fruit and vegetables. Phenyloxidases are defensive enzymes that are activated when plant tissues are ruptured (such as in the peeling of an apple or aubergine) and produce a bacteriostatic film at the point of the injury. Blanching denatures these defensive enzymes before they begin to act. In the production of certain animal foods, the limited dismantling of the tertiary and quaternary structure is achieved with mild heating in order for their internal structures to become accessible to proteolytic enzymes. In cooking, the pressure cooker decreases the cooking time required for meat due to the cumulative action of heat and pressure accelerating the denaturation and hydrolysis of proteins.

5.8.8 Casein: Structure, self-assembly, and adsorption

Casein is perhaps the most important milk protein, and a version of it is produced by practically all mammals. Although these versions have been significantly differentiated by evolution, it is possible to detect common sequences in the individual primary structures. In bovine milk, approximately 80% of the proteins by mass are caseins. Milk caseins can be divided into four categories, called α_{s1}-, α_{s2}-, β-, and κ-casein (a glycoprotein), which occur in the approximate proportions 4:1:4:1.6, respectively. Other caseins exist, such as other α_s-caseins and γ-casein (a product of the breakdown of β-casein), but as these are generally found in very small quantities in milk they will be omitted from our discussion.

The peculiarity of casein lies mainly in the presence of phosphorylated serine, known as phosphoserine. The phosphoric group of phosphoserine binds calcium ions, which in turn bind other phosphoric groups, resulting in the creation of colloidal pockets of calcium and phosphate, which are the main forms in which calcium and phosphorus are stored in the milk.

In α_{s1}- and β-casein, the different monomers are not regularly distributed along the length of the peptide chains; consequently, there are not immediately identifiable self-assembling regions, and no extensive α-helices or β-sheets are formed. However, α_{s2}- and κ-casein appear to have a significant part of their chains formed into secondary structures. These two caseins have two cysteine residues per molecule. Caseins aggregate together to form spherical entities known as casein micelles (not to be confused with the micelles formed by low-molecular-weight surfactants), around 40 to 300 nm in diameter with a total mass of 10^6 to 10^9 Da. A large number of protein chains are found in each micelle. Among the forces that bind the molecules into the micelle are interactions between colloidal calcium phosphate and the proteins, perhaps involving the phosphate group of phosphoserine. This is a very good example not only of the binding of organic molecules by a protein, but also the use of an inorganic substituent for the uniting of two protein molecules, that is, for the development of quaternary structure. There are theories that maintain that there are smaller micelles (known as sub-micelles) dispersed on the inside of every micelle, interconnected with calcium phosphate bridges, while other theories maintain that the real dispersed phase is the colloidal calcium phosphate particles dispersed within a continuous casein phase (i.e., there are no sub-micelles, only bridges connecting pieces of colloidal calcium phosphate).

During cheese-making, phosphoserine-poor κ-casein, which forms the outside of the micelle, is cut off by the action of chymosin, an enzyme found in rennet. The peptide parts that are revealed flocculate under exposure to the continuous phase. This flocculated material can also contain fat, because caseins can be adsorbed to the surfaces of milk fat droplets. The flocculate/precipitate is compacted and processed as required.

5.8.9   Adsorption and self-assembly at an interface: A complex example

In the first half of the nineteenth century, Acherson reported that a solution of proteins (as they would later be called) tends to spontaneously form elastic "skins" around the surface of droplets. This correct observation was extended to become the basis of the description of protein behavior at interfaces.

Interfaces are the real natural environment of a protein. The dominant understanding today is that proteins have been designed by nature more as interface-active agents rather than simple water-soluble molecules. We have already interpreted the folding and the structures of proteins in terms of the physics of interfaces. The most important functions of proteins are surface activities; they include the adsorption of a protein onto a hydrophobic surface (casein onto the envelope of fat droplets in homogenized milk), the adsorption of a lyophilic surface onto a protein (the catalyzed substrate onto an enzyme), the binding of an atom to a protein (iron to hemoglobin), the permanent anchoring of a protein to a lyophobic surface (myosin to connective tissue), and the selective adsorption of a protein to a cell membrane composed of a double layer of surfactants (adsorption of antibodies to cell membranes).

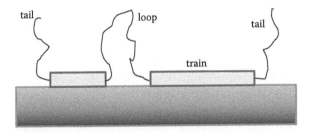

Figure 5.8 Arbitrary representation of two protein chains adsorbed onto a hydrophobic surface by their hydrophobic parts (here represented by rectangles) while the hydrophilic sections extend into the aqueous phase.

The interfaces that mostly interest a chemist or food technologist are clearly those between water and air or fat/oil triglycerides. Let us say that a protein is dissolved (perhaps an odd term for a gigantic molecule with a mass of some hundreds of thousands of Daltons!) in water. We have already said that the macromolecule may not be in its best thermodynamic state because of the incompatibility of the nonpolar (hydrophobic) amino acid residues with water. The proximity of nonpolar amino acids to water significantly increases the free energy of the entire system. The obvious solution is for the nonpolar amino acids to withdraw from the external surface of the protein, which is the reason for gathering the nonpolar amino acids on the inside of the tertiary and quaternary structure. Similarly, when a protein finds itself near an interface that divides the aqueous phase from another, nonpolar phase (e.g., triglyceride or air), the sections of the protein chain that are primarily composed of hydrophobic amino acids will align in order to adsorb to the interface, while those sections that are mainly composed of hydrophilic amino acids will orient themselves toward the aqueous phase, forming tails and loops.

Perfect tails and loops can only be observed in completely free chains without tertiary structure. Although this is not something normally encountered in real proteins, the tails and loops model is a good illustrative example of the conformation that a protein attempts to adopt on adsorption to a surface.

5.8.10   To what extent does the above model apply to the adsorption of a typical spherical protein?

In the above section, it was implied that the intramolecular interactions between the different parts of a protein are very weak. Only if this is the case can the tertiary structure open up easily to spread out on the surface and to form tails and loops. Are these interactions too small? Generally not. During the protein's approach to the surface, enthalpic factors of importance that are competing with one another (leaving aside entropic factors for reasons of simplicity) can be related to:

1. The attraction between individual parts of the protein (e.g., dispersion forces between hydrophobic amino acids, hydrogen bonds, and electrostatic interactions between charged/hydrophilic amino acids)
2. The hydrophobic interactions (dispersion forces) between the non-polar parts of the protein and the surface (these lead to the unfolding of the protein and formation of loops and tails)

The forces involving direct interaction between hydrophobic molecules (i.e., hydrophobic amino acids and oil triglycerides, if adsorption is to occur at an oil–water interface) are relatively weak, as they are not expected to involve hydrogen or ionic interactions. Hence, as the forces in (2) are small, the protein will only unravel a significant part of its tertiary structure if the forces in (1) are weaker by comparison.

Lyophobic repulsions between the surface of a protein and the aqueous phase (one of the basic driving forces for adsorption, let us remember) contribute to the adsorption of the protein to the interface. Despite this, as these lyophobic repulsions are related to the surface of the protein rather than with its interior, they do not contribute to the unfolding of the chain.

What does this mean? Simply, it means that the adsorption of a protein to a surface does not presuppose the loss of its tertiary structure—at least not for a simple transfer from the main bulk of the solution to the interface. The state of a protein that has adsorbed to a surface without losing its tertiary structure is described as a molten globule.

For strictly spherical proteins, the idea that they pass through the molten globule stage is acceptable. A molten globule is essentially a protein that has not lost its tertiary structure to its full extent. Following adsorption, important parts of the tertiary and secondary structures can be lost because the hydrophobic parts leave the core of the protein in order to adsorb to the interface.

5.8.11   Under what conditions does a protein adsorb to a surface, and how easily does it stay adsorbed there?

Let us begin with the adsorption and desorption of a hydrophobic group from an aqueous environment to a hydrophobic surface. The driving force of this adsorption is, as in the case of the folding of hydrophobic parts of the chain, the need to reduce undesirable interactions between the hydrophobic groups and water, which eventually drives the hydrophobic group as far away as possible from the water, that is, to the hydrophobic surface. Desorption will occur when the system is provided with energy at least equal to the energy that was obtained by the system from the adsorption (e.g., from a local increase in temperature). Such temperature fluctuations are very frequent. This means that the adsorption of an individual hydrophobic group to a hydrophobic surface is reversible; there is a dynamic equilibrium between the adsorbed and nonadsorbed form. In surfactants, equilibrium is established at a given concentration and temperature, relating the ratio of adsorbed to nonadsorbed surfactant. This is a dynamic process; molecules continually desorb and re-adsorb in extremely small time scales, so the only thing that can be defined is the statistical probability that a molecule will be adsorbed or desorbed at any given moment. In the current case of a polymeric chain (i.e., a protein) with many adsorbed hydrophobic groups, desorption of the entire macromolecule would imply simultaneous desorption of all adsorbed groups. The probability of such an event is very small, and although some proteins can be removed from an interface with successive washing proteins are generally considered to be adsorbed.

Proteins then, typically macromolecules with many hydrophobic groups, adsorb irreversibly to interfaces. However, proteins can detach from a surface when other surface-active molecules (usually low-molecular-weight surfactants) compete with the already-adsorbed proteins for places on the interface. When a protein has saturated an interface (has covered it completely) the addition of a low molecular weight surfactant (e.g., a detergent or a food-grade emulsifier) can eventually lead to the dislocation of the proteins from the interface and their replacement with surfactant molecules. The main reason for the dislocation of the proteins is the fact that they leave sections of the interface uncovered because of their folded and bulky structure. In this way the adsorption of the surfactants covers the interface more completely and reduces the interface tension. Thus the adsorption of low-molecular-weight surfactants to an interface is thermodynamically favored over that of proteins.

Spherical proteins (e.g., β-lactoglobulin) tend over time to form viscoelastic films at interfaces, probably due to the relatively slow interactions between the adsorbed protein molecules on the interface. Types of peculiar quaternary structure such as these play an important role in the stability and texture of foods such as ice cream and bread.

chapter six

Emulsions and foams

6.1   Colloidal systems

In 1845, Francesco Selmi described those liquid dispersal systems that were not sufficiently clear to be considered homogeneous solutions as "pseudosolutions." In the 1850s, Michael Faraday studied dispersions of gold in water. He observed that the gold, once sedimented out, could not be resuspended. He reached the conclusion that *lyophobic* dispersions are generally unstable entities that are stabilized only kinetically (i.e., not thermodynamically). A little later, in 1861, Thomas Graham defined as colloidal those dispersal systems in which the dispersed particles are sufficiently large (above 1 nm, 10^{-9} m) that they do not display a significant diffusion coefficient in comparison with the small molecules. Today, dispersion systems in which the particles are on the order of nanometers or a few micrometers (1 μm = 10^{-6} m) can generally be considered colloidal. Gels should likewise be regarded as colloids. These, although they are not optically heterogeneous ("pseudosolutions"), are classified as colloidal systems for numerous reasons related to the sub-micrometer particles most of them contain. These give rise to properties such as the scattering of light at invisible wavelengths and their characteristic rheology.

A classification of a colloid should include a description of the *continuous phase* (the phase in which the colloidal particles are dispersed) and the *dispersed phase* (the particles themselves). The description of the continuous phase should include its physical state (solid, liquid, or gas). The corresponding description of the dispersed phase should likewise include its physical state, shape, size, and dispersion of particle sizes (bubbles, droplets, solid particles), and any possible aggregation. Table 6.1 provides a basic guide to the characterization of different colloidal systems.

A characteristic of colloidal systems is the extremely large surface area of the dispersed phase, as the main bulk of the material is usually fragmented into microscopic pieces. As we shall see further on, the free surface of colloids gives them exceptionally interesting physicochemical (and by extension, biological) properties.

A fundamental characteristic of colloids is *Brownian motion*, a continuous movement due to the anisotropic nature of the pressure exerted on their walls. Small fluctuations in temperature around the perimeter

Table 6.1 Selected Examples of Colloidal Systems
Cross-Referenced with the Physical States of Their Two Phases

PHASE		Continuous		
		Solid	Liquid	Gas
Dispersed	Solid	Minerals Salami	Suspensions Mud Plankton Mustard	Smoke
	Liquid	Margarine Butter	Emulsions	Clouds Fog Spray
	Gas	Expanded polystyrene Sponges Bread Whipped cream	Foams	Steam vapors

of a particle lead to the application of a different pressure at each point on its surface (anisotropy). This results in the characteristically irregular Brownian motion.

6.1.1 Emulsions and foams nomenclature

The term *emulsion* defines a colloidal dispersion of a liquid into another liquid. An essential prerequisite for the formulation of an emulsion is, of course, the immiscibility between the two phases. That suggests that the two liquid phases must comprise, for example, two mutually insoluble solvents (e.g., of oil dispersed into water) or a single solvent containing phase-separated regimes, (e.g., of specific aqueous carbohydrate and/or protein solutions that have undergone phase separation as per Section 3.17.1). Sometimes the term "emulsion" is expanded to account for dispersions of an oil or a fat into a gel or into a solid. Under this rationale, a sausage can be classed as an emulsion of oil/fat dispersed into a gel or solid protein matrix, while butter can be considered an emulsion comprised of water dispersed into fat.

Emulsions are named after the two phases that they comprise, that is, the continuous phase and the dispersed phase, the latter normally forming droplets. For example, a cream made of oil dispersed into an aqueous extract is called an oil-in-water emulsion. Inversely, an emollient cream made of water droplets dispersed into an oil phase is called a water-in-oil emulsion. Under this rationale, phase-separated proteins or polysaccharides in aqueous systems such as the ones described earlier are considered water-in-water emulsions.

Because oil-in-water and water-in-oil emulsions consist of two mutually insoluble solvents, work is required for the approach of their molecules, thus creating the oil–water interface between a droplet and the emulsion's continuous phase. Homogenization using the application of some sort of work is required to create the new interface. Creation of the new interface leads to the manifestation of interfacial tension, as discussed in Chapter 4. Droplets will assume, if possible, sphere-like shapes, as the sphere is the geometrical pattern that exposes the minimal possible surface per unit volume. The existence of excess free energy at the interface (due to the work invested in the formation of the interface) will eventually lead to the disintegration of these emulsions into their component phases.

It is possible for an emulsion to be used as the dispersed phase of another emulsion. In that rationale, an oil-in-water emulsion can homogenize in oil to yield an oil-in-water-in-oil emulsion; in a similar pattern, a water-in-oil emulsion can be homogenized with water to yield a water-in-oil-in-water emulsion. In this case, appropriate emulsifiers (i.e., in terms of HLB values) should be used; for example, the formulation of a oil-in-water-in-oil emulsion would require homogenization of oil into water using a hydrophilic emulsifier; that will result in a oil-in-water emulsion, that is an aqueous bulk ("outer") phase. In a second step, homogenization of this emulsion with oil using a hydrophobic emulsifier will result in an oil-in-water-in-oil emulsion. Such complex emulsions tend to be unstable due to osmotic phenomena, that is, an imbalance in the osmotic pressure between the two aqueous phases of a water-in-oil-in-water emulsion could force the inner aqueous droplets to either swell or shrink.

Emulsions comprised of two phases based on the same solvent, e.g., water-in-water emulsions, can be thermodynamically stable (thus "permanent" in some sense). In contrast to oil-in-water or water-in-oil emulsions, formulation of the two phases in a water-in-water emulsion can be spontaneous. In fact, as described in Chapter 3, phase separations transforming a homogeneous solution into a binary system are thermodynamically driven processes. One can reasonably expect that such water-in-water systems present very small, if any, values of interfacial tension between their constituent phases. However, one should not generalize: A number of water-in-water emulsions are not thermodynamically stable, and their life span can be very transient. Water-in-water emulsions originating from phase separations occurring in polysaccharide and/or protein solutions appear to be ubiquitous in the world of food, from the raw materials to their processing and then to food consumption and digestion.

Emulsions tend to consist of droplets ranging from some tenths to some tens of micrometers. The very high curvature of sub-micrometer droplets (1 micrometer (μm) = 10^{-6} m) puts some natural limits on the smallest attainable droplet size. Microemulsions are dispersed systems consisting of "droplets" of even smaller dimensions, sometimes on the order of some tens of nanometers (1 nanometer (nm) = 10^{-9} m). Such systems basically consist of micelles loaded with some molecules of the dispersed phase. Their importance is significant, as they can be utilized as vessels for the microencapsulation of nutrients and/or pharmaceuticals.

Foams are colloidal dispersions of a gaseous phase in a liquid in the form of tiny particles ("bubbles"), although the definition can be expanded to account for non-liquid continuous phases (e.g., ice cream, polystyrene foam) or even to macroscopic rather than colloidal-sized bubbles (beer froth, hand-soap foam, cells in bread and biscuits). Foams are similar in many ways to emulsions: The high interfacial tension of the air–liquid interface is a major contributor to a foam's behavior; surfactant adsorption can stabilize against coalescence, although, as will be discussed, Ostwald ripening-like disproportionation can be a dominant factor in foam (in)stability. Foam drainage, that is, the removal of water from the space between two approaching bubbles, also accounts for the particular behavior of foams. Drainage and gravitational separation can lead to the increase of the volume fraction occupied by the air, resulting in the compression of the bubbles. This is manifested as the familiar polyhedric bubbles that form when we wash our hands with soap or some other detergent.

Many foods are emulsions and foams at the same time; ice cream, for example, is an emulsion comprised of the milk fat dispersed into a partially solidified aqueous phase. Ice cream is usually aerated (air is incorporated into it). Air is trapped in the semi-solid structure and stabilized by means of adsorbed particles, that is, emulsion droplets (which will be discussed later on as a specific case of the so-called Pickering stabilization), while, of course, adsorption of smaller emulsifiers or proteins cannot be ruled out. Whipped cream is also both an emulsion and a foam at the same time. The foam component is instrumental for the specific mouthfeel of such foods.

6.2 *Thermodynamic considerations*

The miscibility of a substance with another substance, like every phenomenon of self-organization, is determined by the combination of an enthalpic factor (here the enthalpy of mixing ΔH_m of the two components) and an entropic factor (here $-T\Delta S_m$, the entropy of mixing ΔS_m at temperature T). Thus, the Gibbs free energy (the sign of which tells us whether the mixing of the two components is spontaneous or not) is given by the formula

$$\Delta G_m = \Delta H_m - T\Delta S_m \tag{6.1}$$

Let us consider two cases of binary liquid systems, A (mutually soluble substances) and B (mutually insoluble substances). The fundamental difference between the two cases lies in the relative contribution of the enthalpic term ΔH_m and the entropic term $T\Delta S_m$ in Equation (6.1). In the case of mutually miscible substances (system A), the enthalpic term ΔH_m is sufficiently large to give its negative sign to Equation (6.1) and ensure that the Gibbs free energy of mixing ΔG_m has a negative value and the mixing is thermodynamically feasible. In the case of two mutually insoluble substances (system B), where ΔH_m is not expected to be strongly negative, only the maximization of the entropic factor $T\Delta S_m$ could lead to a stable system. As we saw in previous chapters, the enthalpy that relates to the creation of free surfaces is very large and positive, and can be larger in absolute value than the entropic term. As a result, colloidal systems can be unstable because of the demands of the surfaces for free energy.

An intermediate case is a system of partially miscible liquids. The free energy of mixing ΔG_m is a function of the molar fraction x of the dispersed substance. One can distinguish three individual regions: At the two extremes of the values of x, where the relative quantity of one of the two components is small, the two individual components mix to form a single phase. In this case, the free energy decreases with a small increase in the substance with the smallest concentration. At intermediate concentrations, between the two minima, the components cannot mix together because a small increase in the component with the smallest concentration will lead to an increase in free energy.

6.3 A brief guide to atom-scale interactions

Up until now our description of colloids has been based on thermodynamics. This section confronts the same problem from the point of view of forces rather than energy. Forces are what determine the kinetic development of colloidal systems, as many colloidal systems are stabilized only kinetically. "Kinetically stable" means that the system remains stable for a reasonably long time, but is not permanent. The time scale of experiments and observations is therefore an important factor when studying emulsions, foams, and similar systems.

To describe the forces that are exerted in a system, we must turn to the study of the forces between atoms and molecules. The analytical description of intermolecular forces is clearly outside the scope of the current work, as the underlying mathematics is extremely complicated. However, a brief overview of the subject is required for completeness.

6.3.1 van der Waals forces

All substances, either in an atomic or molecular state, liquefy and finally solidify at low temperatures. This applies even to highly inert substances such as symmetrical hydrocarbons (e.g., methane) and to noble gases. Even helium, perhaps the most inert of atoms, was solidified under high pressure in 1926. The transformation of a substance from a gas to a liquid and then to a solid state requires the formation of bonds between the constituent atoms/molecules. Let us look more in-depth at the case of helium. The He atom has its outer orbital filled, so logically has no chemical activity; despite this, it does solidify, meaning that forces develop between its atoms. To what can these forces be attributed? They are based on mutual interactions between dipoles that are formed *momentarily* on the helium atoms.

The basic concept is that while the center of density of the electron cloud when averaged over time is located at the center of the nucleus; at a random moment, the two may not correspond. In other words, at a random instant, the center of the negative charges is in a different place to that of the positive charges. Thus, even in helium, such *dipoles* are created momentarily. If such a dipole approaches a nonpolarized atom, it will polarize it, creating an *induced dipole*. Obviously, electrostatic forces will develop between the first (let us say "instantaneous") and the second (let us say "induced") dipole. This interaction is of exceptionally short duration, as it depends on the polarity of the instantaneous dipole. Despite this, it is capable of forming temporary dipole–dipole bonds. These interactions between temporary, instantaneous dipoles are called *London forces*. These are important in the case of nonpolar atoms or symmetrical molecules.

The attractive potential $u_{attr}(r)$ is inversely proportional to the sixth power of the distance between the two molecules with induced dipoles $(u_{attr}(r) = Ar^{-6})$. This means that these interactions are exceptionally short-ranged, and are felt by the molecules only at very small distances. Clearly, this favors tight structures such as those occurring in the interior of proteins. These forces, if acting alone, would obviously lead to a collision between the nuclei. In reality, induced dipole forces become repulsive at still shorter distances. These repulsive forces are considered inversely proportional to the twelfth power of the distance between the atoms $(u_{rep}(r) = Br^{-12})$; that is to say, they are felt at almost atomic-scale distances and are related to the repulsions between two approaching atoms due to the overlap of their electron orbitals. A and B are constants related to the specific interactions between the two atoms. A simple way to quantify the interaction potential between two noncharged particles is given by the Lennard–Jones equation:

$$u(r) = u_{rep}(r) + u_{attr}(r) = A\,r^{-12} - B\,r^{-6} \tag{6.2}$$

Equation (6.2) tells us that during the approach of two atoms, the induced dipole will lead to attractive interactions. These attractions do not end in a collision of the nuclei, since at still smaller distances the interactions become repulsive (dependent on r^{-12}).

Another ensemble of forces applied between two adjacent particles is generated between a permanent dipole and an atom or a nonpolar molecule. These are related to the polarization of the nonpolar molecule by the permanent dipole, in a rationale similar to the one described above for induced dipoles: As a nonpolar molecule approaches a polar one, the electron cloud of the former is either attracted or repelled by the polar molecule, depending on its charge orientation. Such forces are called *Debye forces*.

In the case of two permanent dipoles, it is easier to predict that forces of an electrostatic nature will develop. As one would expect, when two or more polar molecules are in close proximity, they orient themselves so that the electronegative part of one molecule is facing toward the electropositive part of its neighbor and so on, finally resulting in alignment of the molecules. These forces are called *Keesom forces*. All the above forces are collectively known as *van der Waals forces*.

6.3.2   *Hydrogen bonds*

Hydrogen bonds are, as their name implies, related to the presence of hydrogen in a molecule, especially when it is involved in highly polarized bonds. Upon the approach of two molecules, one of which contains a hydrogen-containing bond, say A–H, and one containing another atom, say B, a noncovalent attractive interaction forms them into a complex. The dissociation energy for the products of such interactions is of a magnitude of a few kilojoules per mole (kJ mol^{-1}) up to some tens of kilojoules per mole. This can be taken as an indication of the magnitude of the bonds holding together the two molecules. As a comparison we could mention that the typical energy of a covalent or ionic bond is >200 kJ mol^{-1}. The binding energy of the A–H...B complex increases with the electrostatic potential of A–H and B. That suggests that hydrogen bonds, as a rule of thumb, tend to be stronger for strongly electronegative A and B atoms. Concerning the topography of the bonds, the hydrogen atom lies along the axis between atoms A and B, although it appears that it is relatively easily displaced from that line.

Hydrogen bonds play a major role in stabilizing structures in biological systems, contributing in practically every aspect pertaining to the self-organization of matter, from bringing together distant parts of the primary structure of nucleic acids and proteins so as to fold them into helical secondary structures, assembling polysaccharide supermolecular entities as in the case of cellulose, and, most significantly, forcing a population of tiny molecules such as that of water to remain liquid and not evaporate at ambient temperatures. One should remember that water's "cousin molecule, sulfur dioxide (S_2O, sulfur being under the oxygen on the periodic table) has a melting point of $-72°C$ and a boiling point of $-10°C$. The fact that the lighter hydrogen dioxide (water) remains liquid (and a fairly viscous one) where the heavier sulfur dioxide is a gas should be attributed to the strong hydrogen bonds formed between the water molecules.

6.3.3 Electrostatic interactions

As is apparent from Equation (6.2), dipole forces are attractive at molecular or larger distances, the influence of the repulsive interactions between orbitals $u_{rep}(r) = Br^{-12}$ being significant only on extremely short (atomic) scales. The predominance of the attraction factor at the molecular level would suggest that colloidal bodies would quickly collapse under the influence of van der Waals attractions. However, most colloids collapse gradually rather than instantly. This means that there are forces that partially counterbalance the attractive van der Waals forces. In aqueous systems interactions that usually undertake this task are electrostatic in nature and relate to the repulsive forces between charged surfaces. In general, different surfaces are charged for a variety of reasons, which may include the ionization of surface groups or the adsorption of charged molecules to the surface. For reasons of charge preservation, a cloud of counter-ions surrounds a charged surface. The concentration of counter-ions close to the charged surface leads to the formation of an *electrical double layer* from an immobile surface charge layer and a loose layer of *counter-ions* (Figure 6.1). The repulsive force exerted between two approaching surfaces immersed in water is, in essence, repulsion between the electrical double layers.

On the addition of electrolytes to a solution, the counter-ions congregate at the interface—namely, the physical boundary of their solutions—simply because of their mutual interactions; that is, they try to distance themselves as much as possible from the main bulk of the solution because there the imbalance of the exerted forces pushes them to the interfacial "walls." This enthalpic phenomenon ("enthalpic" here suggesting direct interactions), together with the opposing entropy of mixing (which contributes to the re-dispersion of the ions to the bulk), defines the final concentration profile $\rho(x)$. For reasons of simplicity, we do not account here for enthalpic or entropic contributions involving molecules of the solvent. The distribution of counter-ions will be fluctuating up to a distance d from the surface, while the distribution of counter-ions and the size of the potential in the main bulk of the liquid medium will remain constant. This distance d, which is several atoms thick, is called the *Stern layer*. From a practical point of view, in colloids the interface potential is important the particles move in a fluid. This potential additionally relates to shear flow of fluid tangentially to the particles and is called the *zeta potential*. This is widely used because it is relatively easy to measure experimentally (measurement of mobility in an electrical field).

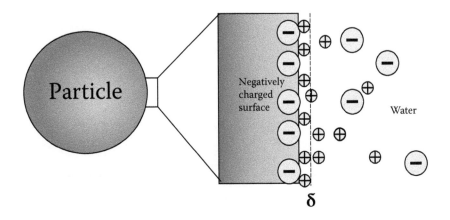

Figure 6.1 Schematic representation of the distribution of ions on the surface of a droplet. A loose layer of counter-ions forms at a distance comparable with the size of the counter-ions.

6.3.4   DLVO theory: Electrostatic stabilization of colloids

The sum of the electrostatic potentials for all distances from the interface is usually positive (repulsive). This occurs when one adds the positive repulsive and negative attractive interactions outside the Stern layer. A stable suspension can form in a bad solvent* phase when the individual particles exert strong electrostatic repulsions on each other in order to overcome the (negative) attractive forces.

Derjaguin and Landau, simultaneously with Vervey and Overbeek, proposed a synthesis of the attractive van der Waals forces and repulsive electrostatic forces in the 1940s. Their summation can satisfactorily explain the stability of some colloidal systems as a consequence of electrostatic repulsions between the charged interfaces of the particles, but also the different forces that have been observed between two surfaces or particles. This theory—the DLVO theory—is named for its four expounders and constitutes the modern basis for explaining the stability of colloidal systems.

According to this theory, two colloidal bodies at distance h have a total interactive potential $u(h)$ equal to the sum of the attractive van der Waals potential $u_A(h)$ plus the repulsive electrostatic potential $u_R(h)$:

$$u(h) = u_A(h) + u_R(h) \qquad (6.3)$$

---

* The term *bad solvent* is mainly used in polymer science. A bad solvent, as opposed to a good solvent, is one that does not satisfactorily dissolve the polymer macromolecules. Here, it is used with the same physicochemical meaning as a system that cannot solvate ("dissolve") a colloidal body.

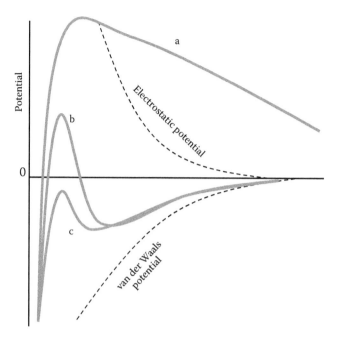

Figure 6.2 Schematic rendering of the DLVO potential as a function of distance for two approaching interfaces. The DLVO potential is presented (solid line) as the sum of the van der Waals and electrostatic potentials (dashed lines). Different possible profiles are presented: (a) the interfaces repel one another strongly, a "stable" colloid, (b) the interfaces come to equilibrium because the energetic barrier is sufficiently large, and (c) the interfaces come into contact rapidly, resulting in coalescence and destruction of the colloid.

Other factors, such as the potential due to the solvent, can also be considered, but are omitted here for the sake of simplicity. Figure 6.2 presents the total DLVO potential as a function of the distance between two approaching surfaces. When the electrostatic forces are strong enough in comparison to the van der Waals interactions, the potential is positive (scenario), preventing the aggregation of two approaching droplets. In scenario (b), during the approach, the attractive force exceeds the repulsive (secondary minimum) up to a point where the resultant forces become positive (repulsive) and form a maximum ("energetic barrier"). This barrier, corresponding to the repulsion between two adjacent electrical double layers, is attributed to the stability (nonaggregation) of colloids by DLVO theory. If the energy barrier is surpassed, the approaching surfaces will aggregate because of the first minimum (which, as previously mentioned, is due to the exponential relationship of the van der Waals forces to the distance). When the electrostatic forces are not very powerful (uncharged interfaces), the energy barrier that ensures stability disappears and the interfaces present strictly attractive forces. The particles in a colloid that have such interfaces flocculate.

DLVO theory can provide useful insights into the stabilization of emulsions: When two droplets in an emulsion or bubbles in a foam, the surfaces of which are covered with a layer of adsorbed surfactant, approach in order to aggregate together, their surfaces repel each other when the distance between them is close to the energy barrier. Because the energy barrier is reduced by the addition of salts due to the neutralization of surface charges, the stability of emulsions and foams is not usually favored by the addition of salts. This is the reason why some detergents lose their activity in hard and salty water.

6.3.5   Solvation interactions

In previous sections we discussed theories regarding the phases of a colloid as continuous media. The calculation of van der Waals forces and the DLVO interactions assume a continuous phase for which only the dielectric constant, the density, and other properties that are independent of its molecular nature are of any relevance. In reality, there is no such thing as a continuous phase, but rather a concentrated dispersion of molecules in a void, usually in constant motion under the influence of strong forces. The distance over which such forces are exerted is on the order of the size of the molecules.

The interactions between the molecules of the continuous phase and any colloidal particles can lead to solvation interactions (leading to repulsion between the dispersed particles) and lyophobic interactions (leading to attraction between dispersed particles). These interactions can cause significant deviations from DLVO theory. The approach of two surfaces (movement of one toward the other) in an aqueous system results in the dehydration of the area between the surfaces (water molecules are displaced from the space between the surfaces). This causes, in turn, an osmotic-like influx of water molecules in the void space, resulting in the creation of a repulsive hydration force.

Hydration interactions and hydrophobic interactions are of exceptional importance in the field of surfactant adsorption to interfaces. Hydrophobic interactions are related to the attraction between two macromolecules or surfaces in the presence of a bad solvent. Lyophobic interactions (of which hydrophobic interactions are a subcategory) have been described either as entropic interaction, or as a general category of van der Waals (London) forces. According to the first approach, the introduction of a macromolecule into water causes changes in the structure of the water, increasing the system's free surface energy due to the break-up of bonds between water molecules and the exposure of more water molecules to the nonpolar phase and often reducing its entropy. To reduce this undesirable change in the energetic content, the polymer molecules aggregate into supermolecular structures that are inaccessible to water. The polymer molecules in aqueous solution fold in order to present their nonpolar (hydrophobic) parts on the inside of the structure, leaving their solvated hydrophilic parts on the outside. The second approach considers that during the coming together of two long (hydrophobic) carbon chains in an aqueous continuous phase, the macromolecules interact preferentially between each other with dispersion (London) forces rather than with the water.

6.3.6 Stereochemical interactions: Excluded volume forces

This category of forces is based on the fact that two material objects cannot occupy the same space at the same time. Although volume exclusion, often called *stereochemical* or *steric*, interactions occur between material bodies of any size, this section concerns large molecular chains. When a polymer molecule is solvated, say in water, it swells due to the hydration of the macromolecular chain: Incorporation of water into the inner parts of the folded structure creates an osmotic pressure that forces the chain to unfold. The stress that the polymer undergoes is analogous to that of a stretched elastic band. At equilibrium, this elastic force is equal and opposite to the osmotic force that caused the swelling.

We saw in Chapter 5 that polymers containing large sequences of hydrophobic monomers tend to adsorb at oil–water interfaces or water–air surfaces. Figure 6.3 gives the density distribution of monomers with distance in the case of a free polymer in solution and a polymer adsorbed onto an interface. It can be determined that the distribution of monomer density changes dramatically on adsorption. The portions of the molecular chain that are adsorbed directly to the interface are called *trains*. The

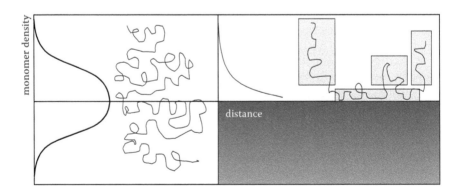

Figure 6.3 Density distribution of monomers with distance for a solvated free macromolecule and for an adsorbed macromolecule (see Figure 5.12). Trains, loops, and tails are highlighted.

parts that do not directly adsorb to the interface tend to be solubilized in the continuous phase and form *loops* and *tails* extending into the continuous phase in order to obtain the greatest possible configuration entropy. In food, the most typical example of macromolecules adsorbing onto interfaces are the proteins. One might intuitively expect the segments that are rich in nonpolar amino acids to be preferentially adsorbed at the interface, while the segments containing mostly polar ones will be oriented toward the aqueous phase. Of course, as mentioned in Chapter 5, this is only a generic rule-of-thumb and has many and marked exceptions.

Let us consider two surfaces covered in polymers (e.g., proteins) approaching one another. At a given distance, as polymer layers start to overlap, a force acts upon them. The nature of this force is complicated and relates to the compression of the polymers onto the surface. This usually leads to repulsion analogous to the repulsion between two electrical double layers. In this process, an additional entropic phenomenon, which has to do with the loss of configurational entropy by the compressed molecules, also comes into play. This occurs because the compression of the polymer and its proximity to other polymers reduces its free movement and consequently the possible configurations that the macromolecular chain can adopt, thereby reducing its entropy. This phenomenon leads to the mechanism known as steric stabilization of colloids and to this is principally attributed the stabilization of emulsions and foams provided by adsorbed macromolecules such as proteins, as well as part of the detergent and foaming activity of many nonionic surfactants: The adsorbed macromolecules on the surface of fat droplets of bubbles of air are repelled by other similar layers and prevent the aggregation of the particles. Some general prerequisites for steric stabilization to occur between two approaching surfaces are (1) the polymers must be firmly attached to the surface, (2) the polymer chains must extend far from the interface, and (3) the surfaces must be fully covered with the polymer.

Another important category of excluded volume interactions is that of *depletion interactions*; these were predicted theoretically in the late 1950s by researchers such as Asakura and Oosawa. These are mutually attractive forces exerted between particles of a colloidal dispersion in a solution of smaller nonadsorbed macromolecules. The force keeping the particles together is directly dependent on the osmotic pressure of the polymer solution in the absence of colloidal bodies (Figure 6.4).

Let us consider two colloidal bodies, say emulsion droplets, of diameter D that are approaching one another in the presence of smaller entities, say macromolecules of diameter d. These smaller particles are expected to exert osmotic pressure on their environment. Concerning the dispersed droplets, this pressure is symmetrically applied normally to the interface, facing toward the droplet's center. As the two moving droplets come close together at an interface-to-interface distance smaller than the diameter

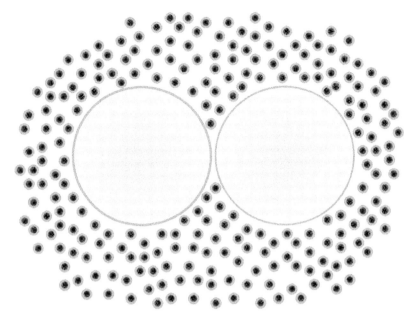

Figure 6.4 Schematic image of depletion flocculation between two droplets that are surrounded by smaller macromolecules. Note that osmotic forces exerted on both sides of the axis connecting the centers of the droplets do not balance, resulting in the droplets remaining united.

of the macromolecules d, the space between them becomes sterically depleted of these macromolecules. The osmotic pressure exerted on every droplet is not zero anymore, because the pressure exerted by the polymer molecules is asymmetrically distributed: On the external sides of the two droplets, osmotic pressure is exerted by the polymer; this, however, is not cancelled out on the opposite sides of the droplet (those at the depleted zone between the two droplets) because this depleted volume does not contain any polymer, and hence does not apply osmotic pressure. As a result, if two spheres come into contact, they do not easily separate.

In the simplest approach, as two rigid spheres of radius r_{drop} approach one another in the presence of non-adsorbing particles of radius r_{pol} the gradient of the polymer concentration between this small volume and the bulk phase gives rise to an osmotic pressure Π that keeps the spheres together.

$$\Delta G_{dep} = -2\pi r_{pol}^2 \, \Pi \left(r_{drop} + \frac{2}{3} r_{pol} \right) \tag{6.4}$$

At a first approximation, the depletion free energy ΔG_{dep} for the interaction between the two droplets is given by Equation (6.4), while the osmotic pressure forcing the spherical droplets together can be approximated by

$$\Pi = \frac{CRT}{M_{pol}}\left(1 + \frac{2C}{\rho}\right) \tag{6.5}$$

where C is the concentration of the macromolecule, and ρ is the density of the macromolecules inducing depletion.

Depletion interactions are responsible for *depletion flocculation*. This can be due to nonadsorbed polymers such as synthetic polymers, polysaccharides or proteins, or micelles of nonadsorbed surfactants. Thus, when phase separation is observed in an emulsion product, this can be due to excess emulsifier that is not adsorbed, or an excess of nonadsorbable hydrocolloids (e.g., carbohydrates such as xanthan or guar gum), which are commonly used as viscosity regulators in such applications. Of course, phase separation in emulsions can also be due to a large number of other causes that do not involve depletion phenomena.

6.4   Emulsification

Emulsions are colloidal dispersions of a liquid in another liquid; foams are colloidal dispersions of a gas in a liquid. A common feature of these two colloidal systems and the physical chemistry of interfaces that was described in the previous chapter is the very large specific surface area presented by the dispersed phase of an emulsion or foam. This means that colloidal systems of this type have a very large quantity of energy trapped as additional free energy in the interface, which is exhibited as high interfacial tension. This occurs because the components of colloidal systems are, in principle, not soluble in each other. According to Equation (6.1), the Gibbs free energy ΔG_m for the mixing will be positive. This means that emulsions and foams are inherently unstable systems, and, thermodynamically speaking, should not exist. However, thermodynamics is only concerned with final states; that is, it states that such colloids will not exist *in the end.*[*] Emulsions and foams can, however, be stabilized *kinetically*; that is, their final breakdown into two separate phases can be delayed, even for very long periods (e.g., years).

In general terms, emulsification is concerned with the breakdown of a phase B into small droplets of colloidal dimensions, their stabilization against recoalescence (which will be expanded upon later), and the

---

[*] Nanoemulsions, which are essentially micelles with some nonpolar molecules solvated on their interior, can be thermodynamically stable entities. The present chapter is not concerned with nanoemulsions, but with common emulsions that have a major application in foods.

dispersion of the formed colloidal droplets in a continuous insoluble phase A. It is the process of producing an emulsion from two immiscible liquids.

As mentioned in Chapter 4, a significant quantity of energy must be provided to a biphasic system for a new free surface to be formed. A colloidal particle (e.g., a droplet of oil in an oil-in-water emulsion) adjusts its shape in order to present the minimum possible free surface area when it is dispersed. The geometric shape that provides the minimum possible surface area per unit volume is the sphere, so when droplets in an emulsion (or bubbles in foam) are formed, they will try to preserve a spherical shape whenever possible. External forces frequently subject the droplets to deformation, which is opposed by Laplace pressure Δp:

$$\Delta p = \gamma \left(\frac{1}{R_1} + \frac{1}{R_2} \right) \tag{6.6}$$

where γ is the interface tension, and R_1 and R_2 are the two basic radii of curvature of a droplet distorted into an ellipsoidal shape.

As is clear from the analogy between the Laplace pressure and the product of interface tension and radius of curvature in Equation (6.6), the Laplace pressure is directly related to the work that is required to form curved rather than flat surfaces. This creates a limiting factor in the formation of very small particles beyond the basic limit that exists because of the large free surface. As we will see later, this leads to Ostwald ripening, that is, the enforced solvation of molecules of the dispersed phase from the droplets into the continuous phase and their migration into droplets of smaller Laplace pressure (i.e., into larger droplets).

For a droplet to be produced from a mass of liquid, or a bubble from a mass of water, initially its deformation toward the dispersed phase is required in order to create the droplet or bubble. This can happen as much in laminar as in turbulent flow. In the case of ideal linear flow, the deformation is created from the shear stresses that are exerted from the surrounding liquid and are directly related to the density of the continuous phase and the differential of velocity to distance. The formation of this differential can consume energy many times that corresponding to the surface tension, and eventually becomes heat. Emulsification, then, is a very energy-consuming process, as are all the processes that concern the provision of energy for the reduction of size (e.g., grinding and milling).

The basis of emulsification lies in the fragmentation of larger droplets into smaller droplets by the provision of mechanical energy. If we consider turbulent rather than linear flow, we see that the fragmentation into droplets can occur more readily. Turbulent flow is described by the

Reynolds number (*Re*), which for a flow of liquid in a pipe is calculated by
the formula

$$\text{Re} = \frac{Du\rho}{\eta} \tag{6.7}$$

where D is the diameter of the pipe, u is the velocity of the fluid, ρ is its
density, and η is the dynamic viscosity of the fluid. For isolated droplets
that move within the fluid, a special Reynolds number can be defined:

$$\text{Re}_{dr} = \frac{2r^2 G \rho_c}{\eta_c} \tag{6.8}$$

where r is the radius of the droplet, G is the differential of velocity with
the distance at a plane perpendicular to the movement of the medium,
ρ_c is the density, and η_c is the viscosity of the continuous phase.

Formation of droplets can also occur under the influence of a laminar field of forces. Such a case is met in the forced passage of a mixture
through a small opening, as happens in the appertures of many industrial and laboratory valve homogenizers. In this situation, the fragmentation of the liquid into droplets takes place in the region of hyperbolic
flow of liquid in the entrance to the valve. The droplet size is directly
dependent on the applied pressure and the resulting interface tension.

Another means of dispersing material into smaller particles is *cavitation*. In a liquid in which the molecules are subjected to an oscillating
vibration, such as from an ultrasound field, pockets of pressure so low
that cavities are formed can occur at the pressure minima of the oscillation. These collapse on a time scale smaller than that of the oscillation
period of the ultrasound, during which process they draw liquid into the
cavity that later remains as a new droplet. This is achieved with the use of
high-energy ultrasound.

Newly formed droplets can recoalesce (Figure 6.5). Recoalescence is
the merging of smaller, freshly produced droplets into a larger one—a
direct reversal of the emulsification process. The same volume of liquid
can recoalesce many times during the same passage through the homog-
enizer. In order for a droplet to be stabilized, for example during the
desorption of a stain from fabric during washing, or during the homog-
enizing of milk into smaller fat droplets, surfactants (detergents or proteins, respectively, in the above cases) must diffuse from the continuous
aqueous phase near the newly created fat droplet–water interface and
then adsorb onto it. The covering of the interface with surfactants creates

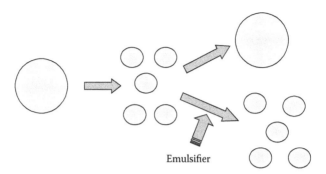

Emulsifier

Figure 6.5 Fragmentation of a large drop into individual droplets. The droplets will re-aggregate unless sufficient emulsifier adsorbs to the interface.

an electrostatic or steric protective layer that renders the droplet water-soluble, inhibiting its re-adsorption to the fabric or its recoalescence with other fat droplets.[*]

Mention must also be made of stabilization by particles (Pickering). During this stabilization, non-surfactant particles (e.g., fat crystals, silica) adsorb onto the fat–water interface, the adsorption not necessarily being driven by lyophobic interactions. The resulting emulsions have a notable resistance to coalescence and particular mechanical properties.

6.4.1   Detergents: The archetypal emulsifiers

The basic mode of action of detergents is that they facilitate the wetting of droplets of the contaminant, which is an oily phase and therefore insoluble in water. Consider an oily droplet on a substrate such as cloth. The nonpolar contaminant has, as a rule, a greater chemical relationship to the relatively nonpolar cloth than to the polar water (given that in the simplest case, the washing will be carried out with water). The contact (wetting) angle φ is acute (see Section 4.4), which means that thermodynamically the contaminant droplet "prefers" to stay attached to the cloth rather than be solvated in the water. For the droplet of oil to be removed from the cloth substrate, the surface tension must be reduced. The surfactants achieve exactly that, resulting in an increase in the wetting angle φ and the gradual departure of the droplet from the cloth substrate. Usually, a large contact angle does not ensure the automatic removal of the droplet from the cloth because the droplet remains attached with a very small

---

[*] Naturally, the provision of mechanical energy (rubbing, mixing) creates an air–water mixture. Surfactants also adsorb to the air–water interface, stabilizing the foam. For this reason, foam is observed during washing with detergents. The foam is often perceived subconsciously by the consumer as evidence of the efficacy of the detergent, despite the fact that it binds valuable surfactant to the wrong interface!

surface on the cloth. At that point, the final detachment is achieved with the provision of mechanical energy (e.g., rubbing, the rotation of the washing machine). Another approach is the use of a washing medium with a greater chemical affinity for the oily droplet than the droplet has for the cloth (e.g., an organic solvent or supercritical carbon dioxide rather than water). In any case, the common method is the lifting of the contaminant droplet from the substrate (cloth) into the continuous phase (washing liquid).

6.5 Foaming

Foaming is the dispersion of a gaseous phase into a liquid one. As mentioned earlier, sometimes the term "foam" also encompasses dispersions of gases into solids: for example, bread, biscuit or bone, and dispersions of gases into liquids/gels that result in an overall gel or gel-like structure (e.g., ice cream or whipped cream). As with emulsification, the prime requisite for foaming is the break-up of the phase to be dispersed (here, gas) and its dispersion into the continuous phase. If a material capable of protecting against coalescence, such as a surfactant, protein, or particles capable of imparting Pickering stabilization, covers the interface via diffusion and adsorption, the foam can be protected against bubble coalescence. However, foams are still prone to disproportionation, which is discussed later.

One of the most common ways to incorporate gas into a liquid is to vigorously stir the surface of the latter, thus driving air into the bulk of the liquid phase, subjecting the newly created air bubbles to strong shear fields and thereby breaking them up into smaller bubbles. In other cases, such as beer or carbonated drinks, a saturated solution releases gas after mild shaking. In still other cases, such as whipped cream-type ready foams, compressed air is released into the liquid matrix (water containing emulsifiers, proteins, or Pickering stabilizers, among other things), resulting in the instant formation of a foam. Another scenario relates to the release of gas from microorganisms, and its entrapment into a mechanically strong protein matrix, as happens in the case of holes in cheese.

As air has a much lower density than liquids encountered in foods, bubbles in a typical food or drink foam will tend to separate to the top of the liquid, a phenomenon familiar to the reader as the formation of froth at the top of a glass of beer. When the upward movement of bubbles can be arrested due to the very high viscosity or due to the elasticity/solid character of the dispersed phase, as in the case of bread, cheese, or ice cream, bubbles can become trapped in the structure, thus resulting in a more or less homogeneous bubble distribution in the food matrix.

Common experience suggests that, in food, foams such as the froth of beer, whipped cream, and even bread, the volume fraction φ of the dispersed phase tends to be large, usually greater than 0.5. For a mono-modal size distribution of spheres (i.e., all spheres being of the same size), a maximum packing is achieved in principle at a sphere's volume fraction $\varphi \approx 0.7405$. However, random packing of undeformed spheres results at lower volume fractions, with $\varphi \approx 0.64$ being a realistic value. In the case of spheres of very diverse sizes, although theory suggests that specific populations of undeformed spheres can achieve a volume fraction much higher than 0.74, this number is close to a realistic upper limit in many cases. This is due to the fact that the effective volume fraction includes not only the volume occupied by gas (or liquid in the case of emulsions), but also the surface layer that moves along with the spherical bubble. This layer can be the protein layer, or the electrical double layer in the case of charged surfactants.

Attractive forces between the bubbles of a concentrated foam can result in the coming together of the bubbles and their truncation and subsequent loss of spherical shape as each bubble tries to accommodate itself between its neighbors. The reader should be familiar with such shapes made of water in the presence of soap. Such phenomena are relevant to *drainage*, which is the removal of water from the space between two adjacent droplets, resulting in the formation of a thin film between them and subsequently the collapse of the foam.

6.6   *Light scattering from colloids*

Beer foam is white; it is composed, however, of transparent air and yellowish-transparent beer. Frothy waves splashing on a beach, or soap foam in our hands are also formed from transparent air and water; still they are not as transparent as their constituents. The same can be said for the whiteness of milk. The above phenomena are due to the fact that light is scattered when traversing colloidal systems. Scattering is the phenomenon that occurs when light traverses a boundary separating two materials of different *refractive index n*, and results in the alteration of its propagation direction. For a given material, the refractive index is defined as the ratio of the wavelength of light in vacuum over that of the wavelength in the material in question. The refractive index is dependent on λ, with a usual reference wavelength being 589 nm.

In the case of emulsion droplets or foam bubbles, the main factors determining the extent of light scattering are the size of the particle, the wavelength of the incident radiation, and the refractive index of the particle and that of the continuous phase. When the scattering particles are considerably smaller than the wavelength, incident light tends to scatter isotropically, that is, the intensity of scattered light is the same at all angles. As the particle size becomes comparable to that of the incident light, scattering starts to become more complex, and the intensity varies with the angle. When the scattering object becomes considerably larger than the incident light, scattering becomes much stronger at smaller angles and less pronounced at larger ones. A number of theories have been developed as to correlate the pattern of scattering at a given wavelength to the particle size and to the refractive indexes; such approaches are used for the determination of the particle size distribution of colloids.

Visible light typically ranges in wavelength between ~390 and ~750 nm; that is very roughly half a micrometer. Emulsions typically found in foods contain droplets on the order of a few tenths of a micrometer to some tens of micrometers; foams in general contain somewhat larger particles. That means that light will scatter anisotropically from such particles, while photons of each wavelength will scatter in a different pattern. If we add up the effect of multiple scattering, that is, a photon scattering from a single particle will scatter again from another particle and so on, it is easy to understand that a sequence of photons of various wavelengths being emitted from a body and passing through a typical emulsion will scatter in such a way that their relative positions will be disrupted, exiting the emulsion in a more or less chaotic sequence. Mixing up these colors produces the typical white appearance of most emulsions and foams containing very small bubbles. Of course, in many cases, stronger scattering of certain colors can preferentially confer hues of particular colors, for example, blueish hues in particular cases.

6.7   *Destabilization of emulsions and foams*

It is better to speak of *destabilization* of emulsions and foams rather than *stability* because, as we have said, it is rather difficult to define an emulsion or foam as something thermodynamically stable, at least when the discussion pertains to oil-in-water or water-in-oil emulsions or typical foams. However, we can study the *manner* and the *rate* with which an emulsion loses its colloidal properties and gradually transforms into a macroscopically heterogeneous mixture of two or more insoluble components.

Here, the previously mentioned concept of kinetic stability is addressed: The viability of an emulsion or foam is determined by the extent to which and for how long it is stable in the face of a series of particular processes that will eventually destroy the colloid. The technological meaning of the stabilization of an emulsion of foam is that is should remain stable for a clearly defined time period (e.g., until the expiry date of the product).

The mechanisms that can lead an emulsion (or a foam, the reasoning is the same) to destabilization can be classified into five general categories (Figure 6.6), which are described in the following subsections.

6.7.1   *Gravitational separation: Creaming*

We are all familiar with the bubbles that rise to the surface of a carbonated soft drink. Their movement is due to the fact that the carbon dioxide gas has a lower density than water, and thus rises to the top. For the same reason, beer froth collects at the top and not the bottom of the glass. In the case of sewage sludge at biological treatment plants, the heavier sludge particles sediment downward out of the lighter water. The separation of the phases of a colloidal system that occurs due to the difference in their densities is called *gravitational separation*. In an oil-in-water emulsion, the droplets, usually of lower density than the water, rise to the top. In emulsion technology, this is known as *creaming*, from the separation of milk fat into two layers under the influence of gravity, one pressed to the surface that contains a high proportion of fat (35% to 65% v/v are realistic values), the *cream*, and one deficient in fat, the *serum*. At intermediate stages of creaming, an intermediate layer can be observed that has a fat content equal or close to the initial fat content. This is the bulk of the dispersed fat that is gradually rising to the top.

Creaming, and gravitational separation more generally, does not in itself destroy an emulsion, foam, or other colloid, but produces a layer of densely packed droplets/particles clearly separable from the continuous phase that has been emptied of colloidal particles (the serum). In the concentrated layer of droplets or bubbles, it is easier for flocculation/aggregation or coalescence to occur and for the colloid to be destroyed, as we will see later.

Gravitational separation is the result of competition between gravitational movement and diffusion, the latter tending to redisperse the droplets that have accumulated at a particular point in the system. Diffusion is more important for small droplets, while larger droplets tend to be influenced in practice only by gravitational forces, resulting in complete

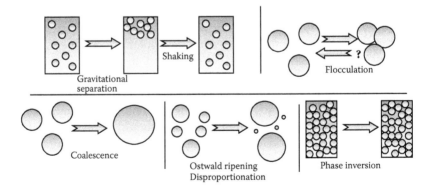

Figure 6.6 The five basic mechanisms by which emulsions destabilize.

separation into "cream" and "whey." Colloids with polydisperse particle sizes (particles with populations of dissimilar sizes) can exhibit both behaviors and develop a large-droplets-rich cream and small-droplets-rich serum.

In a first approach, the velocity v of a spherical particle of radius r in a gravitational field modified for frictional drag is given by the Stokes equation:

$$v = \frac{2r^2 (\rho_o - \rho) g}{9\eta_o} \qquad (6.9)$$

where g is acceleration due to gravity, η_o is the viscosity of the continuous phase, and ρ_o and ρ are the density of the continuous phase and of the particle, respectively. From the proportionality of velocity with the square of the radius, it is obvious that the larger particles will move much faster than the smaller under the influence of gravity. Consequently, an emulsion is more stable against creaming the smaller its particles. Similarly, with a larger difference in density, creaming occurs more quickly, as well as with an increase in acceleration due to gravity (e.g., under centrifugation). This is why centrifugation is used in the preparation of skimmed milk. As would be expected, high viscosity (as in dense solutions) inhibits creaming, as the upward movement of droplets or bubbles is hindered by a viscous continuous phase (and the presence of other similar particles). By manipulating the parameters, one can use the Stokes equation as a rough guide to control the rate of gravitational separation up to a point.

The Stokes equation, although very useful as a guide, ignores many phenomena that make the mathematical description of gravitational separation more complicated. Some of the additional parameters that one would have to take into account in a satisfactory description of the phenomenon of, for example, creaming are the following[*]:

- "Slip" between the droplet and the surrounding liquid of the continuous phase. This makes it easier for the droplet to move, increasing the speed of gravitational separation. In the case where there is an adsorbed layer of surfactant (such as a detergent or a protein) around a fat droplet, the extent of slip can depend on the viscoelasticity (see Chapter 7) of the adsorbed layer.
- The deformability of the droplet in shear flow. This is directly connected to the interface viscoelasticity and the viscosity of the droplet.
- Hydrodynamic phenomena due to diffusive flow of liquid in the opposite direction to that of the droplet. These interactions become more important in cases where there is a large dispersed phase content due to the increased probability of two droplets approaching one another.

[*] For a thorough discussion, see, i.e., Dickinson and Stainsby (1982); McClements (2004).

- Flocculation. Compact, isolated flocs tend to move more quickly due to their higher hydrodynamic radius.
- Large variation in particle sizes. The larger droplets move quickly, drawing the smaller particles along with them.
- Charged particles that interact can present a lower velocity of gravitational movement. This happens because the charges inhibit the particles from approaching one another and exploiting the uplift that is caused by the movement of another particle.
- High viscosity (see Chapter 7) of the continuous phase (e.g., in the case where droplets move in a solution of macromolecules). The high viscosity of the continuous phase is opposed to the tendency of the droplets to move during creaming.
- Brownian motion must be accounted for in unflocculated systems, especially for small droplets.
- A very high concentration of droplets can inhibit or completely stop creaming.
- Extended flocculation can create a structure-spanning network of droplets. This network can act as a scaffold, thereby inhibiting the upward movement of its component flocs.

6.7.2   Aggregation and flocculation

Flocculation, aggregation, and agglomeration are terms that can be confused. In general, the individual branches of the sciences of colloids and surfactants do not agree on the terminology. All these terms essentially suggest the coming together of particles, liquid droplets, or gas bubbles. As a result of this clustering, structures can be formed, comprised of a few to very large numbers of particles. The term *flocculation* is particularly used in emulsion technology to suggest the aggregation of droplets to give three-dimensional structures, without the droplets coalescing (this term means the unification of two droplets into one and is discussed later).

Aggregation/flocculation can be of major importance, as it leads to drastic alterations of the structure of the emulsions and foams. The grouping of droplets or bubbles into clusters affects the other mechanisms of destabilization, such as coalescence or creaming, due to the proximity of the droplets. As the clustering of particles generates far larger entities, Brownian motion can be arrested, while the flocculated colloids, such as yogurt, tend to have a very different macroscopic appearance and texture from their nonflocculated predecessors.

The coming together of droplets can be due to a number of different factors. From a chemical point of view and with greater interest in emulsifier technology, a number of mechanisms can be proposed for the interpretation of aggregation/flocculation:

- Blocking of the charges of the electrical double layer that provides electrostatic stabilization.
- Alteration of the continuous phase composition. A typical example is the substitution of a good solvent for a bad one, for example, the addition of alcohol to the water of emulsions stabilized with a very hydrophilic surfactant, or the acidification of milk, which causes proteins to lose their solubility and aggregate.
- Irreversible bridging flocculation, in which a polymer with two or more hydrophobic regions adsorbs separately to two or more particles, thereby connecting them.
- Bridging by divalent or trivalent cations (i.e., Ca^{++}). This similar mechanism plays an important role in the flocculation of charged polymers and, consequently, in the flocculation of particles that are coated with them.
- Depletion flocculation. This particular mechanism has been presented analytically in Section 6.3.6.

Flocculated droplets can drastically alter the behavior of an emulsion. It was already mentioned in the discussion on creaming that isolated flocs having a larger radius than individual droplets cream faster as per the Stokes Equation (6.16). On the other hand, very large flocs can form networks spanning throughout the volume occupied by the emulsion. In such cases, a flocculated structure can have a significant impact on an emulsion's stability and overall behavior, as it acts as a scaffold for the entire emulsion. This can arrest the process of creaming and cause the emulsion to gel. Emulsion gels are not unusual in complex foods.

Droplets flocculated in open structures tend to spread and occupy more space than the same droplets flocculated in closed, compact structures. One of the most widely used approaches for the characterization of the openness or compactness of flocculated structures in colloid science is the examination of their fractal character. Let us consider a series of droplets flocculated so as to form a one-dimensional filament. Such a structure would require each droplet to link with two other droplets, and so on. This single-dimensional entity by definition is said to have a *fractal dimensionality D* equal to 1. If we imagine droplets forming a flat surface—that is, a two-dimensional structure—we have a *fractal dimensionality D* equal to 2. If we imagine a filament with intersected "bonds" with other droplets (as to create branched structures), a structure results that has a fractal dimensionality somewhere between 1 (a nonbranched rod) and 2 (a compactly built surface). Using the same rationale, a compact assembly of droplets in a three-dimensional space can be said to have a fractal dimensionality close to 3. If we imagine a looser structure involving branching of droplets in three-dimensional space, that would have a fractal dimensionality between 2 and 3. This is an elegant way of quantifying the form of the flocs under a single parameter.

6.7.3   Coalescence

As mentioned previously, the provision of energy for the formation of an emulsion or foam has as a target the creation of new liquid or gaseous particles of colloidal dimensions. The energy provides the free surface and necessary curvature for the creation of small droplets or bubbles to the material for dispersal. This stored energy can eventually be liberated on the coalescence of two particles into a larger one, so as to reduce the free energy of the interface and the associated curvature. When two droplets or, better yet, two surfaces approach one another in the absence of surfactant, the thin interface layers aggregate, forming a bridge ("neck") with material from the dispersed phase that unites the two droplets or bubbles. The neck then broadens, completely merging the two droplets or bubbles.

Let us follow a typical coalescence closely: Suppose that two droplets are approaching one another. Their curved surfaces deform, losing their spherical symmetry. The shortest distance between two spheres is on the imaginary line that connects their centers. The coalescence in the absence of surfactant begins there. In the case that molecules of surfactant are adsorbed onto the interface, a thin lamella is formed. Local oscillations (transverse waves) in the pressure to which the lamella is subjected can cause it to rupture, which is potentially counteracted by the Gibbs–Marangoni effect, due to which surfactant molecules, if present, tend toward the critical point of the collapsing lamella. Surfactants of larger molecular mass, such as proteins that are too immobile to diffuse effectively to the critical point, stabilize the surface by quenching the intense oscillations in local pressure that cause the collapse of the surface. They also inhibit the approach of the two surfaces by means of osmotic interactions between the adsorbed macromolecular layers, as discussed on steric stabilization in Section 6.3.6.

Coalescence between two liquid droplets is related to interfacial film rupture and will lead to the formation of a single, spherical droplet (steric provisions allowing). If the droplets, however, are not liquid (as in the case of solidified fat), crystals can protrude out of the droplet, and such crystals can merge with similar crystals protruding from neighboring droplets. Although their dispersed phase components have merged, the two solid droplets will not readily lose their shape. This phenomenon is called *partial coalescence*, and is usual in cold emulsions with solidified dispersed phases, that is, in ice cream.

6.7.4 Phase inversion

In phase inversion, the dispersed phase of a biphasic system becomes continuous, and vice versa; for example, a water-in-oil emulsion becomes an oil-in-water emulsion. It is reasonable to assume that phase inversion would occur more readily in an emulsion with a high volume fraction of the initially dispersed phase. A characteristic example of phase inversion is the bursting of soap bubbles: A very large volume fraction of air dispersed in a thin membrane of water is destroyed instantly, the air becoming the continuous phase and the small amount of water in the membranes transforming into microscopic droplets scattered in the air. Another typical example is the manufacture of butter: Cream (an oil-in-water emulsion of very high oil volume fraction) is mechanically agitated to force its inversion into butter, a water-in-oil emulsion.

The most notable characteristic of phase inversion is the sudden change in physical properties (viscosity, conductivity). In emulsions, the typical value of dispersed volume fraction at which the physical properties change increases with the concentration of emulsifier and the change can be significantly retarded. The process itself is attributed to a moment of mass coalescence of droplets or bubbles, usually of a high oil volume fraction, in which the droplets/bubbles (which because of their high volume fraction are very close to one another) coalesce en masse and form the new continuous phase, with the former continuous phase trapped as isolated droplets.

Phase inversion depends on the nature of the surfactant and on the HLB value, and likewise on the temperature. At a critical *phase inversion temperature* (PIT), characteristic for every emulsion and concentration, the emulsion turns from an oil-in-water to a water-in-oil emulsion, or vice versa. As the PIT is closely related to the surfactant (emulsifier), the PIT value can be used along with the HLB, cmc, and Krafft point, among other parameters, to help select the suitable emulsifier for each application.

6.7.5 Disproportionation and Ostwald ripening

From the above one could maintain that as long as the interface layers remain intact and thus prevent coalescence, two droplets or bubbles will remain as two discrete entities. In reality, however, particularly in foams, the number of bubbles decreases while their average size increases without significant coalescence being observed. What is happening? In these systems, the process known as *disproportionation* prevails, which is connected to the broader physicochemical process known as *Ostwald ripening*. The latter is a phenomenon in which the molecules that comprise the dispersed phase particles (e.g., air molecules in a foam) dissolve in the continuous phase (e.g., water) and then diffuse until they are resolvated in another particle (bubble). During this process, the molecules that are found in the smaller particles (smaller bubbles) tend to transfer to the continuous phase more readily than those contained within larger particles. Thus, the larger particles slowly grow at the expense of the smaller ones.

At the end of the 1950s, Lifshitz and Slezov and, independently, Wagner described this process from a physical point of view, producing the theory now known from their initials as LSW theory. Attempting to interpret the aging of alloys (metal-in-metal dispersions), they arrived at the LSW equation. A description of it is

$$V = \frac{dr^3}{dt} = f\left(v_1^0, \gamma, S_\infty, D\right) \tag{6.10}$$

In this equation, V is the rate of the process, which is equal to the differential of the radius of the particle r cubed (i.e., volume in terms of dimensions) with time t. The parameters on which this rate depends are the molar volume of the dispersed phase v_1^0, the interface tension γ, the solubility of the dispersed phase molecules in the continuous phase S_∞, the diffusion coefficient D of the molecules of the dispersed phase in the continuous phase, the temperature T, and the universal gas constant R.

As can be seen in Equation (6.10), the main characteristic of Ostwald ripening is the linear relationship between changes in r^3 and in t. This linearity is very useful because it allows us to check whether the breakdown of an emulsion or foam is due to coalescence of the droplets or bubbles or due to Ostwald ripening.

The above is most important technologically because the strategy for protecting the emulsion or foam can be different in the two cases. To protect the system from coalescence, the interface layer must be reinforced with more effective or a greater quantity of surfactant (attention to the HLB, cmc, and Krafft point!). As is evident from the LSW equation, retardation of destabilization due to Ostwald ripening can be achieved by (among others) reducing the solubility and the diffusion coefficient of the dispersed phase (modification of the dispersed or continuous phase), reducing the volume fraction of the dispersed phase, and modifying or controlling the temperature and interface tension. The choice of surfactant (foaming agent, emulsifier) is important because it influences the surface tension, diffusion coefficient, and solubility (molecules solvated in the continuous phase between micelles). Furthermore, although not clearly indicated in Equation (6.10), the development of a mechanically strong and elastic interface layer (e.g., interlinked proteins) can, under certain conditions, offer stability because it resists the continual shrinking of the droplets (see Figure 6.7).

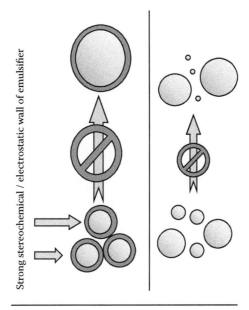

Strong stereochemical / electrostatic wall of emulsifier

Reducing the solubility of the lipid in the aqueous phase

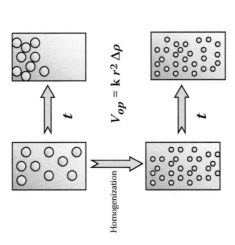

$$V_{op} = \mathbf{k}\, r^2\, \Delta\rho$$

Homogenization

Figure 6.7 Strategies for the prevention of creaming, coalescence, and Ostwald ripening.

chapter seven

Rheology

7.1 Does everything flow?

Rheology is the science that studies the flow of material. By "flow" we mean the reorganization of the components of a system under the influence of an external force. Most objects are subjected to external force fields (which can certainly be powerful in many cases), and the stresses produced can deform them (i.e., their components will lose their spacing and positions relative to each other). Can all objects be deformed? Potentially yes, as long as the appropriate force is applied to them for a sufficient period of time. In reality, it is the *perception of time* that limits the *perception of flow*: The basic concept of flow, as understandable by the observer, is that of observable change (e.g., the movement of liquid in a tube) throughout the time for which the observation takes place. The above definition places the basic limit on human understanding of flow: Can we define flow only as changes that we can observe directly? No, we cannot. For example, we can easily observe the flow of water out of a glass: The time we can spend observing the flow is more than enough for a significant dislocation of the liquid to occur. The situation gets more complicated if we have a container with cold walnut praline or mastic gum dessert: Inverting the container may not bring about a directly observable flow. Cold praline and mastic gum dessert can flow over a time span on the order of tens of minutes or more, while our observation is reasonably restricted to some tens of seconds. If we leave the container upside down and come back in half an hour, we will see that the praline has altered its shape and moved downward.

Consequently, from a purely practical point of view, the ratio of the relaxation time t_s that a phenomenon (e.g., the translocation of a quantity of liquid along the length of a channel) takes to the time t_o for which it is observed is important. This ratio is known as the Deborah number (De).[*]

Here we attempt a first approach to the issue of what a "fluid" is, as opposed to what a "solid" is. Further on we will see that these terms are both inadequate; for a correct description of the material world, it would be better to use the terms "viscous" and "elastic." The aim of the following

---

[*] The Deborah number is named after the Old Testament prophetess who sang that "the mountains will melt in front of you, Lord." Here we should also recall Herakleitos' dictum "Πάντα ρει" ("everything flows").

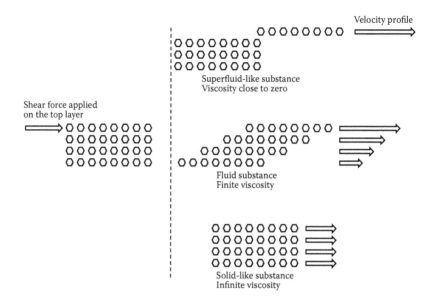

Figure 7.1 Momentum transfer to lower layers after application of a shear stress. Note the basic differences between ideal gas, liquid, and solid; consider its effect on the manifestation of viscosity.

passage is to help the reader make this transition as easily as possible. The simplest way is the gradual introduction of the concept of elasticity of solids, in juxtaposition with the Newtonian flow and viscosity of liquids and gases. To continue, the states of plastic or nonplastic flow will be described, finishing with the hybrid term *viscoelasticity*, which as a rule covers the majority of edible and cosmetic preparations.[*]

Here we briefly discuss the basic theme of the paragraphs to follow. Let us imagine a material body comprised of molecules organized in layers (Figure 7.1). If one applies a force to the top layer, with a direction parallel to the surface (this will subsequently be called a *shear force*), three basic scenarios can occur, depending on the magnitude of the forces existing between the molecules comprising this body. The first scenario assumes that the attractive forces between the component molecules range from very weak to negligible. In that case, the top layer, to which the force is applied, will accelerate and move, leaving behind the layers underneath it. The careful reader may link this behavior to that of

---

[*] Here, the difference between the *rheology* and the texture of a product must be stressed. Texture is a combined sensory stimulus that can relate not only to the rheology, but also to the appearance (pouring of a sauce from a bottle), the sound (the breaking of a crunchy food in the mouth), and also with the subjective impression of every individual.

ideal gases discussed in Chapters 1 and 2 and in reality it characterizes the so-called superfluids. The second scenario assumes that attractive forces do exist between the molecules, linking to some extent the molecules of two adjacent layers. In that case, momentum will be transferred to the layer under the first one, which, in turn, will transfer part of its momentum to the layer underneath it, and so on. This will result in a collective movement of these layers, albeit with decreasing velocity as we move from the top to the lower layers. The gradient of velocity with distance from the top layer (to which the driving force is applied) will be correlated to *viscosity*. This is a typical behavior of liquids; the reader might consider here correlating the development of intermolecular forces and viscosity to the transition from gases to liquids discussed in Chapter 2. The third scenario assumes very strong bonding between the subsequent layers of molecules. This means that the applied force will be dissipated on equal terms to all layers underneath the first layer, resulting in an equal velocity in every layer. This way, the form of the material body will be preserved, as all of its components maintain their relative positions in the structure while it is moved under the influence of the force applied to the top layer. This, of course, is a solid body. Here, the reader might consider the effect of strengthening the intermolecular forces in the transition from liquid to solid, as discussed in Chapter 2.

7.2   Elastic behavior: Hooke's law

The basic behavior of a solid body during the application of stress was described by Hooke. According to this description, the relationship between the Force F that is applied to a body and the induced small deformation x is linear:

$$F = -kx \qquad (7.1)$$

The proportionality constant k describes the reaction of an ideal spring to deformation parallel to its axis (linear deformation). The proportionality constant is represented by k and referred to as *Hooke's constant*.

In more complex cases, a relation between stress and strain is used. In cases of linear deformation (tension or compression) the constant of interest of the material remains constant, it is represented by E and called the *elastic modulus*, or *Young's modulus*, and is equal to the tensile stress divided by the tensile strain. When a material undergoes compression or stretching during which a change in volume takes place because of pressure from every side of the material, such as hydrostatic pressure for

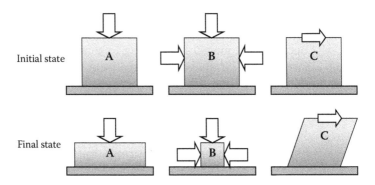

Figure 7.2 Consequences of applying linear (A, B) and tangential stress (C). In B, the volume of the body is changed by the application of stress, but not in cases A and C.

example, the constant of interest is usually represented by K and referred to as the *bulk modulus.*[*] For shearing tension, the proportionality constant is represented by G and called the *shear modulus* G, or μ (Figure 7.2).

In the particular case of linear tension that leads to a change in the length but not in the volume (case A in Figure 7.2), the *Poisson ratio* ν can be defined as the negative ratio of the relative change in size (strain) of a body transverse to the applied force to the relative change in size along the axis of the applied force. The above apply for small relative changes.

Hooke's law was first formulated to describe the mechanical behavior of springs: As a spring is pulled away from its relaxed position, a force proportional to the distance that the spring has moved is applied to it with a direction opposite to the deforming force. If one lets the spring go, this force will cause the spring to revert back to its initial, relaxed position. When this is attained, the force will become zero. The cause of this reversibility is the sum of the cohesive forces between the structural units of the material. The application of external tension causes the dislocation of these structural units (e.g., of the crystals in a metallic lattices and salts or the molecules of organic materials). Provided that the deformation is small, the work applied for the deformation of the system will be stored as an enthalpic component in the bonds holding the system together. When the applied force is withdrawn, the bonds will rearrange the system to its

---

[*] This quantity is equal to the negative produce of the volume times the derivative of pressure over volume

$$K = -V \frac{dp}{dV}$$

initial form while releasing the above-mentioned enthalpic component in the form of kinetic energy turned into heat.

In general, elasticity applies when (1) the material is homogeneous and isotropic, meaning that its properties are the same at every point and in every direction; and (2) the deformations are small. When the deformations become larger, the interactions between the structural elements of the material weaken and the forces that resist the deformation disappear. Under the influence of the external force, the crystals or molecules are reordered in space in order for the forces between them to neutralize the externally applied stress. This process is known as *relaxation*. Thus, an elastic material such as a piece of rubber or a spring can be stretched beyond a limit up to which the deformation is proportional to the tension, with the result that the deformation is no longer reversible.

7.3 Viscous behavior: Newtonian flow

The concept of viscosity was described for the first time by Newton (1642–1727). Let us consider a liquid between two plates of area A and separated by a distance x. If a force F is applied to the upper surface, it creates a tangential stress $\tau = F/A$, expressed in units of Pa ($N\ m^{-2}$). This means that the movement of the upper layer drags the lower layer, because the molecules of the individual layers develop bonds between them. As discussed at the beginning of this chapter, the transmission of movement from the upper to the lower layer is obviously greater in materials with stronger intermolecular bonds. This change in velocity with distance from the upper layer is called the *shear rate*.

Now suppose that the upper surface, to which is initially applied a force F, moves with velocity v (Figure 7.3). The shear rate $\acute{y}$ can be defined as the derivative of the velocity to the vertical distance ($\acute{y} = dv/dx$). The units of this quantity are seconds^{-1}. In this system, the viscosity η can be defined as the coefficient of proportionality between stress and shear rate:

$$\sigma = \eta\, \acute{y} \tag{7.2}$$

From the above it appears that the viscosity is the resistance to flow of a fluid. For a given shear stress, high viscosities correlate with low shear rates. The units of viscosity according to Equation (7.2) are Pascal-seconds (Pa s).

Equation (7.2) predicts the ratio of applied stress to the *rate* of deformation/shearing, *and not* to the deformation/shearing itself as with elastic materials. The difference lies in the fact that the cohesive forces

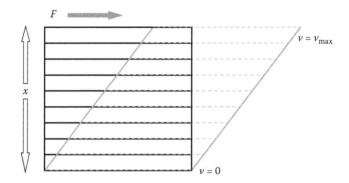

Figure 7.3 Typical Newtonian flow. Note the linear relationship between velocity and depth.

between structural components of a fluid (molecules, ions, or crystals) are not able to muster a reaction proportional to the deformation, so whatever change occurs in the relative positions of the molecules or ions is irreversible.

This equation can describe the flow of gases, pure liquids of small molecular mass, dilute solutions, and very dilute colloidal systems. All of the preceding fluids, for which the shear rate is proportional to the applied stress, are called *Newtonian fluids*. Non-Newtonian fluids (e.g., molten polymers, dense emulsions) can exhibit Newtonian behavior over a limited range of shear rates, but deviate from this behavior usually at very high or very low shear rate values.

7.4   Non-Newtonian flow

7.4.1   Time-independent non-Newtonian flow

It is usual in incompressible real materials, including foods, for the stress and shear rate to not correlate linearly ($\eta \neq$ constant). Depending on the value of η, a series of states may be observed (Figure 7.4):

- The first scenario occurs when η is a constant, that is, $\tau = \eta\, \dot{y}$. This is the basic expression of Newtonian flow.
- The second scenario occurs when η decreases with increasing strain rate: These conditions define a material that starts to flow with the application of a minimal stress, while its viscosity decreases with increasing stress. This behavior is called *pseudoplasticity* or *shear thinning*, and is perhaps the most common non-Newtonian rheology. At very high stresses, the viscosity stabilizes at very low values. Thick emulsions/creams typically belong in this category.
- The third scenario occurs when η increases with increasing strain rate. Fluids exhibiting this behavior are called *dilatant* or *shear*

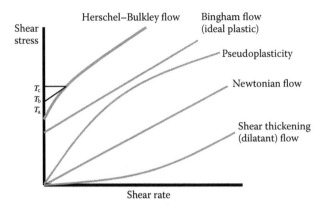

Figure 7.4 Different formulae for the flow of fluid and semifluid materials.

thickening fluids, and include thick solutions/suspensions of carbo-hydrates such as starch pastes (e.g., dough) and thick dispersions of solid in liquid as in many types of mud and sludge (e.g., quicksand).

- The fourth scenario occurs when flow does not begin until a criti-cal stress has been reached. In its simplest form, this phenomenon is described by the equation $\tau = k + \eta\, \acute{y}$. Materials following this behav-ior are called *Bingham plastics* (or *ideal plastics*). Flow begins as soon as the applied stress surpasses a threshold k, which is known as the *yield stress*. At low stresses, the plastic material behaves as a purely elastic material. Once the stress passes the value of k, the material behaves as a Newtonian fluid (the viscosity remains constant at all shear rates).

- In a more generic case, the equation $\tau = k + \eta\, \acute{y}^n$, known as the *Herschel–Bulkley model*, improves the ideal Bingham model for plastic materials. Once again, the plastic material behaves as a purely elas-tic material, storing the energy provided at low stresses. In this case, three different flow rates can be defined: The real flow rate τ_a that is the abscissa to the stress axis, the projection of the equation $\tau = k + \eta\, \acute{y}^n$ onto the stress axis τ_b (which resembles the yield stress of the Bingham model), and the abscissa of the point where the above behavior starts to the stress axis τ_c. The exponent n is called the *flow behavior index*. Because it is an exponent, it is dimensionless. The index n does not have the same value at all of the possible regions of shear rate, as it is not a stable characteristic of the fluid. Despite this, the index n tends to be used as a constant for very narrow ranges of val-ues of sheer stress.

Plastic flow in general is described by equations of the type $\tau = k + f(\acute{y})$. Their basic characteristic, as discussed for the Bingham plastics, is that the materials behave as elastic solids until a critical threshold k.

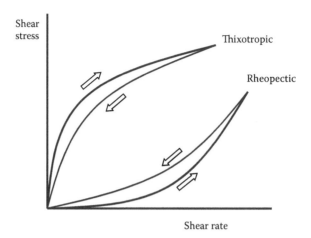

Figure 7.5 Flow curves of thixotropic and rheopectic materials. Note the hysteresis during the return of the shear rate to zero.

7.4.2   Time-dependent non-Newtonian flow

The above cases describe rheological behavior without taking into account the changes that occur in the system over time. In reality, especially in solutions of macromolecules, redistributions and rearrangements in space of the dispersed molecules, changes in the structure and shape of the molecules, and other phenomena such as loss of solvent to evaporation can lead to a gradual change in the rheology of a material. The situation becomes more complicated because the application of shear stress to a solution of macromolecules can itself influence the arrangement and structure of the macromolecules.

The rheological behavior of such systems is a function as much of the *history* of the material as of the shear rate. Even if the former generally relates to inelastic materials (basically to solids), it is important in the behavior of gels, emulsions, and other colloidal systems in general. The time dependency of rheological behavior can be described as a reduction in viscosity over time (*thixotropy*) or as an increase (*rheopecty*, also known as rheopexy).

The characteristic property of thixotropic fluids is the reduction in viscosity following the extensive application of shear stress. Hysteresis can be observed in thixotropic systems when the shear rate increases and then decreases (see Figure 7.5). Starch paste can be a typical thixotropic fluid.

The opposite behavior to thixotropy is rheopecty: An increase in viscosity under the influence of shear stress and constant temperature is observed. Thick, pasty dispersions of metals in liquid display rheopectic behavior. As with thixotropy, hysteresis can be observed in cycles of increasing and decreasing shear stress.

Under ideal conditions, the viscosity of rheopectic and thixotropic materials tends to return gradually to its initial value once the stress has been removed. This ideal situation generally applies in practice to rheopectic materials, but less so in the case of thixotropic materials. With the lifting of stress, even after a long time has passed, the final viscosities can be different: in thixotropic materials lower than the initial value, in rheopectic materials somewhat higher. The former is encountered mostly in macromolecular gels with the irreversible loosening of the intermolecular network.

7.5   Complex rheological behaviors

The above categorization does not imply that the systematic classification of the behavior of a material into one of the above categories is easy. The examples that were given as typical of a particular rheological behavior can present a different rheology under another range of shear stresses. This is particularly common in systems such as lightly flocculated emulsions of foods or cosmetics with a dispersed phase of intermediate volume fraction (φ ~0.2–0.5).

A typical example is the rheology of a typical emulsion, which, as is evident from Figure 7.6, is Newtonian at very low shear stresses (the viscosity is independent of the applied stress) or even elastic. Beyond a certain stress limit, the viscosity of the dispersion decreases dramatically and it exhibits pseudoplastic behavior.* After a second stress value boundary, the behavior became Newtonian, this time with a lower viscosity value.

7.5.1   Application of non-Newtonian flow: Rheology of emulsions and foams

Cream is a typical example of a "real" as opposed to an "ideal" material. Everyday goods such as food, detergent, pharmaceuticals, cosmetics, and hygiene products are, in reality, complex colloidal systems. Changes in the above systems are expressed as sudden changes in their viscosities. A good teaching example is to follow the rheology of a colloid, such as the previously mentioned cream, as a function of the volume fraction

---

* This change generally correlates to the breakdown of flocculates under the influence of tangential stress, and this value serves as an indicator of the strength of the flocculation.

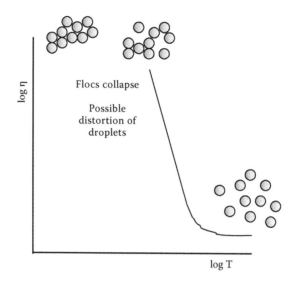

Figure 7.6 Change in viscosity of a flocculated emulsion over a broad range of shear stresses.

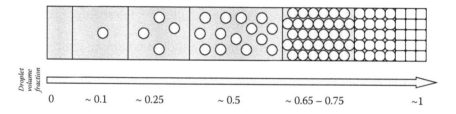

Figure 7.7 Changes in structure of an emulsion as a function of the volume fraction φ of the dispersed phase droplets.

of the dispersed phase φ. As is apparent in Figure 7.7, at a small volume fraction φ, the distances between the dispersed droplets are large. In this case, the main resistance to deformation caused by shear stress is the viscosity of the continuous phase. As the possibility of the droplets colliding with each other is small, their presence generally does not impede the translocation of the structural elements of the system. However, movement of the particles within the continuous phase results in loss of energy through heat, which becomes more important with increasing volume fraction of the dispersed phase. In addition, an increase in the volume fraction of the dispersed phase increases the probability of collision between droplets. This leads to resistance to flow, increasing the viscosity of the emulsion beyond that of the continuous phase: The dispersed phase starts to actively contribute to the overall viscosity. The first approximation of the new viscosity is given by Einstein's equation[*]:

$$\eta = \eta_0 (1 + 2.5\varphi) \qquad\qquad (5.21 \text{ and } 7.3)$$

where η is the real viscosity and η_0 is the viscosity of the continuous phase.

Einstein's equation predicts a linear relationship between the viscosity and the volume fraction φ. In reality, however, as the volume fraction φ rises, the viscosity does increase, but in a nonlinear manner. In dairy products, the nonlinear relationship between measured viscosity and fat content is well known. Thus, there are deviations from Einstein's equation that can be characterized as the ideal behavior of colloidal systems. Where are these deviations found? The equation presupposes strictly spherical and completely rigid dispersed particles of relatively small volume fraction. In reality, the droplets of an emulsion can be deformed. One of the usual causes of deformation is shear stresses experienced by the larger droplets.

What happens in dense emulsions and foams? As the volume fraction of the dispersed phase increases, the contribution of the dispersed droplets to the overall viscosity can become more important than that of the continuous phase. The linear relationship between viscosity and volume fraction is lost, while the rheology becomes non-Newtonian. As the volume fraction of the dispersed phase increases, the rheology becomes pseudoplastic, while at very high values of φ, the system behaves as a plastic and exhibits viscoelasticity. This occurs because, as the droplets come into contact, the flow of liquid in the continuous phase is disrupted, which in turn disrupts the flow of the whole macroscopic liquid system itself. In very concentrated emulsions (e.g., mayonnaise or some cosmetic creams), droplets are also distorted due to compression by neighboring droplets, and movement of one against the other results in very large amounts of the kinetic energy provided to the system being dissipated as heat.

If the droplets are flocculated and form a continuous network throughout the colloid, they present a further obstacle to induced deformation. In this case, the viscosity is high and in some cases independent of the stress. When the shear stress is sufficiently large so as to overcome the attractive forces between its constituent particles, the flocculate breaks down into isolated droplets or smaller flocculates ("flocculate collapse"). The system then exhibits Newtonian rheology.

---

[*] It is indicative of the contribution of Einstein to the science of colloids that the "Einstein equation" in physical chemistry usually refers to the formula $\eta = \eta_0 (1 + 2.5\varphi)$, and not to $E = mc^2$!

7.6   How does a gel flow? (Viscoelasticity)

The phenomena of thixotropy and rheopecty discussed in Section 7.5 evinces clearly that the history of a system plays a major role in its future rheological behavior. The concept of memory is important in the rheology of a solid. As we have seen, an elastic material has a "memory" of its initial shape and, following a small deformation, it will return to this original shape. If the deformation is large or lasts for a long time, such as to permanently disrupt its internal structure, then the material undergoes plastic deformation (i.e., it does not return to its original condition).

In real systems, the deformation generally depends not only on the applied stress, but also on the duration of its application: In most biological systems (and consequently in foods), time-dependent rheological phenomena exist. If a material undergoes only a small deformation, then the ratio of stress to deformation is independent of the degree of deformation. This behavior is called *linear viscoelasticity*. Within this range of deformations, the Boltzmann superposition principle applies, according to which the consequences of the mechanical history of a system are cumulative. In other words, if a material undergoes a series of stresses at different moments, the deformation at a given moment will be equal to the sum of the deformations that it has undergone altogether, as if all the individual stresses had been exerted simultaneously. Experimental methods for measuring viscoelasticity are largely based on this principle. The motivation behind this analysis is the existence of materials that routinely appear in food and cosmetic technology for which the question "is it fluid or solid?" cannot be immediately answered nor quantified with that which has been described so far. The aim of the techniques described below is precisely this quantification of the solid or liquid state of a viscoelastic material such as a paste, a gum, a gel, or a dense emulsion.

7.7   Methods for determining viscoelasticity

7.7.1   Creep

Creep is measured by the application of stress and the observation of the subsequent deformation for a given time period. The results are expressed as the *creep compliance J* for the time-dependent deformation $e(t)$:

$$J(t) = \frac{e(t)}{\sigma} \qquad (7.4)$$

Recovery after removal of stress is another parameter of importance. An ideal elastic body to which stress σ is applied will deform to a particular value J_0. On the lifting of the stress, the deformation will cease and the material will try to return to its original dimensions.

7.7.2 Relaxation

Relaxation tests are essentially the inverse of the creep tests. The material under study is deformed to a given degree, while the stress applied by the material as a reaction to this deformation is recorded with time. The initial stress opposed to the deformation is related to the elasticity of the material. As the deformation is prolonged, the component molecules of the materials can rearrange themselves in the space available to them in order to dissipate the external force field to which they are subjected. This process is called *relaxation* and is a significant mechanism for the measurement of the behavior of concentrated colloids and plastics.

7.7.3 Dynamic measurements: Oscillation

Creep and relaxation tests usually concern the phenomenological behavior of a material during observation times of tens of minutes to hours (very broadly and with many exceptions, 10^2 s to 10^4 s). Dynamic measurements are complementary to the above, as they concern the study of materials over observation times on the order of seconds or less. This is important because the different molecules that make up a material behave differently over different time scales (recall the Deborah number).

The basic dividing lines between "solids" and "liquids" are, as mentioned in the introduction:

1. (a) In a solid, the *deformation* is linear to the applied stress.
   (b) In a fluid, the *rate of deformation* is linear to the applied stress ("viscous flow").
2. (a) In solids, complete reversibility of deformation is (ideally) observed once the stress is removed ("elasticity").
   (b) In fluids, the shape that the stress imposes is adopted permanently, even once the stress is removed ("viscous character").

Consider that in the region of linear viscoelasticity (at small deformation values where, let us remember, the ratio of stress to deformation is independent of the deformation), a variable sinusoidal deformation* γ is applied with maximum value γ_0:

$$\gamma = \gamma_0 \sin(\omega t) \tag{7.5}$$

For an angular velocity ω, the application of oscillating stress is equivalent to a number of tests of time $t = \omega^{-1}$. During the application of the

---

* Equivalent experiments can take place with controlled stress rather than deformation. Here we use the example of controlled deformation because, in that way, the sample is protected from excessive deformation.

deformation, the material is expected to react with stress τ. An ideal elastic body according to Hooke will react against the deformation

$$\tau = \tau_0 \sin(\omega t) \tag{7.6}$$

This stress $\tau = f(\omega t)$ can be presented as the sum of two components perpendicular to one another, one synchronous with the applied deformation and a second displaced by 90° from the first ($\Delta(\omega t) = 90°$):

$$\tau = \gamma_0 [G'(\omega) \sin(\omega t) + G''(\omega) \cos(\omega t)] \tag{7.7}$$

The parameters G′ and G″ are called the *storage modulus* and *loss modulus*, respectively. In combination, they give the *complex modulus* G*, with the storage modulus as the real part and the loss modulus as the imaginary part:

$$G^* = G' + iG'' \tag{7.8}$$

The storage modulus G′ is a useful measure of the extent to which a material is elastic ("solid"), in contrast to the loss modulus that indicates the extent to which a material is viscous ("fluid"). With this reasoning we can quantify the solid or liquid characteristics of substances such as gels and creams using the values of the two moduli. The correlation of the two parameters can provide useful information about changes in rheology during processes such as the setting of yogurt illustrated in Figure 7.8, in which the two moduli are plotted as a function of time at a frequency of 1 Hz. The sudden increase in the storage modulus at around 7,800 s (≈130 min) indicates the transition of the milk to yogurt.

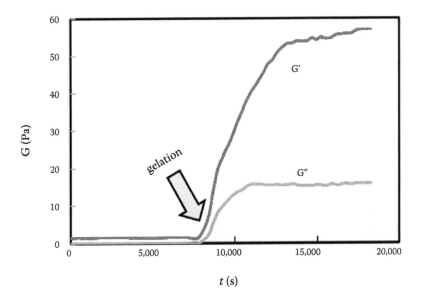

Figure 7.8 The setting of yogurt as monitored by an oscillation study (frequency 1 Hz) in a U-tube rheometer in the laboratory for the study of the physical and chemical parameters of foods in the Department of Food Technology, ATEI Thessaloniki. (From data provided by G. Vlachavas ATEI Thessaloniki, Greece.)

chapter eight

Elements of chemical kinetics

8.1 Diamonds are forever?

If one were to calculate the free energy of the reaction $C_{(diamond)} \rightarrow C_{(graphite)}$, it would be discovered, somewhat unexpectedly, that the thermodynamically preferred form at normal temperature and pressure is graphite. This means that diamond is thermodynamically unstable, and for this reason graphite—and not diamond—is used as the most stable form of elemental carbon in Hess' law (see Chapter 2). Why then does diamond not spontaneously convert to graphite? The reason is that the tetravalent interconnection of the carbon atoms in diamond is very strong (diamond is scored at 10 on the Mohs scale and is the hardest substance known). This means that there is enormous *kinetic* difficulty in reordering the carbon atoms into a graphite matrix, although there is a thermodynamic tendency for the transformation and, as a result, the conversion of diamond to graphite practically never occurs spontaneously.

With similar reasoning, thermodynamics tells us that emulsions and foams should not exist because they require huge quantities of energy bound up in interfaces. However, this author, despite thermodynamics, has a glass of milk next to him. It could *eventually* (e.g., in a few months) separate into a layer of fat over a layer of water if it remains undisturbed, but *for the moment* it is stable. As discussed in Chapter 6, before an emulsion or foam is destroyed by physical laws (thermodynamics), it can remain stable for a period of time (kinetics).

Chemical thermodynamics, with which we were concerned in the first chapters, studies whether a reaction can take place and up to what point. In the first chapters, we clarified and emphasized the fact that thermodynamics is not concerned with the concept of time: It is interested only in the initial and final state of a system, ignoring whether the changes that it predicts will take place in a second or in a century. *Chemical kinetics* complements thermodynamics at this point, studying how quickly a reaction will take place.

In this chapter the more basic (in the judgment of the author) elements of the kinetics of chemical reactions are presented. The correlation of the law of reactions with their stoichiometry and the mathematical principles of zero-, first-, pseudo-first-, second-, and higher-order reactions are described. To finish, elements of catalysis in general and the kinetics of enzyme activity are presented, due to their importance in food chemistry.

8.2   *Concerning velocity*

In physics the change of a quantity with time is called *velocity*. Thus, the velocity of motion or simply velocity of an object v is the change in its position s over time t:

$$v = \frac{ds}{dt} \tag{8.1}$$

In physical chemistry, the term *velocity* (or rate) refers to the quantification of the progress of a chemical reaction with time, while chemical kinetics is the branch of physical chemistry that measures the velocity (or rate) of reactions. This velocity obviously relates to a chemical reaction in which products are created and increase in concentration and reactants are consumed and decrease in concentration. Thus, the velocity (or rate) v of a reaction as to a particular material (reactant or product, let us call it A) is given by the relationship:

$$v = \frac{dC_A}{dt} \text{ or } v = \frac{d[A]}{dt} \tag{8.2}$$

According to this concept the velocity for the reactants is negative and for the products it is positive (their concentration increases with time).

8.3   *Reaction laws*

Let us consider a reaction in which 1 mol A and 2 mol B react to form 3 mol C:

$$A + 2B \rightarrow 3C$$

Similarly, let us consider that, despite what was said in Chapter 2 about chemical equilibrium, that the reaction proceeds completely to product C. The changes over time of the concentrations [A], [B], and [C] of the three components A, B, and C, respectively, are

$$v_A = -\frac{d[A]}{dt} \tag{8.3}$$

$$v_B = -\frac{d[B]}{dt} \tag{8.4}$$

$$v_C = \frac{d[C]}{dt} \tag{8.5}$$

with a negative sign for the reactants A and B (which decrease in concentration) and a positive sign for the product (which increases in concentration).

From the stoichiometry of the reaction it is apparent that for every 1 mol A that is consumed, 2 mol B are consumed in order to produce 3 mol C. Therefore the velocity with which substance A is consumed is half that of B and equal to a third of the rate of formation of C:

$$v_A = -\frac{d[A]}{dt} = -\frac{1}{2}\frac{d[B]}{dt} = +\frac{1}{3}\frac{d[C]}{dt} \tag{8.6}$$

Likewise the velocity of the consumption of B and the formation of C are equal to

$$v_B = -\frac{d[B]}{dt} = -2\frac{d[A]}{dt} = +\frac{2}{3}\frac{d[C]}{dt} \tag{8.7}$$

$$v_C = +\frac{d[C]}{dt} = -3\frac{d[A]}{dt} = -\frac{3}{2}\frac{d[B]}{dt} \tag{8.8}$$

The above are adequate to describe the rate of the disappearance or formation of *each reactant or product separately.* However, since v_A, v_B, and v_C differ from each other, this approach cannot give us a unique number that describes the course of the equation A + 2B → 3C over time. Obviously, the reaction coefficient (1, 2, and 3 for A, B, and C, respectively) relates directly to the reaction velocity. Consequently, the coefficients of a chemical equation are directly related to the mathematical description of velocity.

In practice, the reactions are classified on the basis of the sum of the exponents of the equation

$$v = k\,[A]^a[B]^b[C]^c \dots [X]^x \tag{8.9}$$

The sum of the exponents a, b, c, ..., x is called the *order of the reaction.* The exponents are usually (but not always!*) the coefficients of the reactants of the reaction of interest:

$$aA + bB + cC + \dots + xX \rightarrow \text{Products}$$

---

* In general, it is not advisable to predict the rate expression from the stoichiometric equation alone. A host of factors can influence its rate, such as a reaction being the sum of multiple reactions, or its progress being inhibited by its reagents.

Note that, according to Equation (8.9), for a single reaction, the velocity theoretically depends only on the reactants and not on the products. The overall equation that describes the concentration of the products and the reactants as a function of time is called the *velocity/rate law* for the equation in question.

For the quantitative description of the development of a chemical reaction with time the concept of *half-life* $t_{1/2}$ is useful. This time is equal to the time required for the concentration of the reactants to reduce by half. This parameter is more practical than the time that is necessary for the reactants to disappear completely since, in accordance with the law of chemical equilibrium, this would probably never happen. To continue, a brief description and classification of reactions on the basis of their reaction order will be given.

8.4   *Zero-order reactions*

Into this category are placed those reactions whose velocity is independent of the concentration of the reactants, i.e., their rate is constant. In this case we consider that the total of the indices in Equation (8.9) is zero (hence the term *zero-order reaction*), so the velocity is constant at every concentration ($v = k$ for every value of [A], [B], etc.).

Let us say that substance A breaks down to substance B, with the stoichiometry given by the equation A → bB. The velocity, which it is reminded does not depend on the concentration of the reactants, is given by Equation (8.3), with the clarification that v_A is constant

$$v_A \equiv k = -\frac{d[A]}{dt} \tag{8.10}$$

while the concentration of substance A at any moment $[A]_t$ is given by integrating Equation (8.9) between the initial concentration $[A]_0$ and $[A]_t$:

$$-d[A] = kdt \Rightarrow -\int_{[A]_0}^{[A]_t} d[A] = \int_0^t kdt \tag{8.11}$$

$$\Rightarrow -([A]_t - [A]_0) = kt \Rightarrow [A]_t = -kt + [A]_0$$

Clearly, a graph of the reactant concentration against time for a zero-order reaction is a straight line that intersects the concentration axis at $[A]_0$ while the reaction constant k is given by the gradient of the line (Figure 8.1).

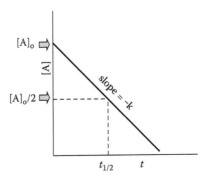

Figure 8.1 Graph of the course of a zero-order reaction.

The half-life of a zero-order reaction is taken from Equation (8.11), replacing $[A]_t$ with $[A]_o/2$:

$$\frac{[A]_o}{2} = -kt_{1/2} + [A]_o \Rightarrow t_{1/2} = \frac{[A]_o}{2k} \tag{8.12}$$

8.5 First-order reactions

In this category are placed reactions of which the velocity is proportional to the first power of the concentration of a reactant; that is to say, Equation (8.9) becomes

$$v = k\,[A] \tag{8.13}$$

The concentration of reactants at time t is given by integration between the initial concentration $[A]_o$ and $[A]_t$:

$$-\frac{d[A]}{dt} = k[A] \Rightarrow -\frac{d[A]}{[A]} = kdt \Rightarrow -\int_{[A]_o}^{[A]_t} \frac{d[A]}{[A]} = \int_0^t kdt \tag{8.14}$$

$$\Rightarrow \ln[A]_t = -kt + \ln[A]_o$$

According to Equation (8.14), the graph of the natural logarithm of the reactant concentration against time is a straight line that intersects the concentration axis at $\ln[A]_o$, while the reaction constant k is given by the gradient of the line (Figure 8.2).

With similar logic to that in Equation (8.12), the half-life for first-order reactants is given by the equation

$$t_{1/2} = \frac{\ln(2)}{k} \tag{8.15}$$

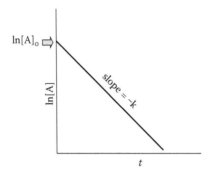

Figure 8.2 Graph of the course of a first-order reaction.

The above assume a basic equation of the form A → products. However, equations of the type A + B → products can be encountered in which the concentration of B remains practically constant (e.g., when [B] ≫ [A]), with the result that the reaction velocity is determined only by [A]. In this case, [B] is calculated as part of the reaction constant, so the velocity is given by a law of the first degree

$$\upsilon = k[A][B] = (k[B])[A] = k'[A] \qquad (8.16)$$

and is called a *pseudo-first-order reaction*.

8.5.1　*Inversion of sucrose*

One of the most important processes in food technology is the splitting of sucrose into D-glucose and D-fructose under acidic conditions.

Even though the hydrolysis reaction seems to be bimolecular, in reality it follows the first-order equations since it is a pseudo-first-order reaction (as there is much more water than sucrose).

The reaction can be written as:

$$C_{12}H_{22}O_{11} + H_2O \rightarrow C_6H_{12}O_6 + C_6H_{12}O_6$$

Sucrose + water → D-glucose + D-fructose

The course of the reaction can be easily observed as follows: Sucrose rotates the plane of polarized light to the right, as does D-glucose, but D-fructose rotates the plane of polarized light to the left. Since the specific rotating ability of D-fructose is greater in absolute value than that of D-glucose, solutions of hydrolyzing sucrose change from being dextrorotatory to levorotatory during the course of the reaction.

The reaction can be followed with a simple polarimeter. During the polarimetry the plane rotation angle of the polarized light a for a solution of a sugar of concentration [S], molecular mass M, and specific rotation $[a]$, measured in a polarimeter with a light path of l dm is given by

$$a = [a]\frac{[S]lM}{1000} \tag{8.17a}$$

The starting solution of pure sucrose of concentration $[S]_0$ and molecular mass 342 g mol⁻¹ shows a rotation of a_0:

$$a_o = [a]_s \frac{342[S]_o l}{1000} \tag{8.17b}$$

Consider that after time of hydrolysis t the concentration of sucrose will reduce by x. From the stoichiometry of the inversion reaction, x will also be the concentration of glucose and fructose (both of molecular mass 180 g mol⁻¹) at the same moment.

The rotation of polarized light by a degrees is a cumulative property; that is to say the sum of the contributions of the individual components to the angle of rotation (here, sucrose, $[a]_S$, glucose, $[a]_G$, and fructose, $[a]_F$), gives the final angle of rotation at time t:

$$a_t = \frac{l}{1000}\left[342[a]_s([S]_0 - x) + 180[a]_F x + 180[a]_G x\right] =$$

$$= \frac{l}{1000}\left[342[a]s[S]_0 + x\left(-342[a]_s + 180[a]_F + 180[a]_G\right)\right] = \tag{8.18}$$

$$= \frac{l}{1000}\left(342[a]_s[S]_0 + xK\right)$$

Combining the final form of Equation (8.18) with Equation (8.17), we obtain:

$$x = \frac{\alpha_t - a_o}{l}\frac{1000}{K} \tag{8.19}$$

At the end of the hydrolysis $x = [S]_0$. Consequently the final rotation a_∞ will be given by the relationship

$$a_\infty = \frac{180[S]_o l}{1000}\left([a]_G + [a]_F\right) \tag{8.20}$$

Combining Equation (8.18) for $x = [S]_0$ with Equation (8.20) we find that

$$\alpha_\infty - a_o = \frac{l}{1000}[S]_o K \Rightarrow [S]_o = \frac{\alpha_\infty - a_o}{l}\frac{1000}{K} \tag{8.21}$$

and with the same logic

$$[S] = [S]_o - x = \frac{\alpha_\infty - a_t}{l}\frac{1000}{K} \tag{8.22}$$

Assuming the reaction to have first-order kinetics (Equation (8.14)):

$$\ln[S]_t = -kt + \ln[S]_o \Rightarrow k = \frac{1}{t}\ln\frac{[S]_o}{[S]_t} = \frac{1}{t}\ln\frac{\alpha_o - a_\infty}{\alpha_t - a_\infty} \tag{8.23}$$

Equation (8.23) can be used to quantify the progress of the hydrolysis of sucrose by measuring the optical rotation of the solution. For example, following the hydrolysis at 25°C of a sucrose solution of initial optical rotation 50.20°, at 180 min 9.27°, and at the end of the reaction −15.30°, by substituting the values into Equation (8.23) we obtain

$$k = \frac{1}{180}\ln\frac{50.20 - (-15.30)}{9.27 - (-15.30)} = 5.4 \cdot 10^{-3}\,\text{min}^{-1} \tag{8.24}$$

It is obviously good practice to check the value of k using other time points before assigning a final value. The primary data can be adapted firstly to check the reaction law, and then to determine the reaction constant.

8.6   Second- and higher-order reactions

The velocity of a second-order reaction depends on the concentration of the reactants where the sum of a + b in Equation (8.9) is 2. This means that for the reactions

$$A + B \rightarrow \text{Products}$$

and

$$2A \rightarrow \text{Products}$$

the velocity law can be given by one of the equations

$$v = k[A]^2 \tag{8.25a}$$

or

$$v = k[A][B] \tag{8.25b}$$

Integration of the preceding equations (Equation 8.25) leads, respectively, to the expressions:

$$\frac{1}{[A]_t} = \frac{1}{[A]_o} + kt \tag{8.26}$$

or

$$\frac{[A]_t}{[B]_t} = \frac{[A]_o}{[B]_o} e^{([A]_o - [B]_o)kt} \tag{8.27}$$

The graph that derives from Equation (8.26) for the correlation of concentration with time is a straight line ($1/[A]$ against t) that intersects the concentration axis at $1/[A]_o$, while the reaction constant k is given by the line gradient.

The half-life is given by the equation

$$t_{1/2} = \frac{1}{k[A]_o} \tag{8.28}$$

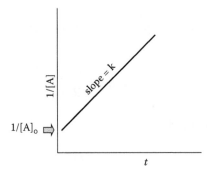

Figure 8.3 Graph of the course of a second-order reaction.

By the same reasoning, we can define nth-order reactions (where n is the sum of the exponents in Equation (8.9)). In this general case ($n > 2$) velocity is given by the relationship

$$\frac{1}{[A]_t^{n-1}} = \frac{1}{[A]_o^{n-1}} + (n-1)kt \tag{8.29}$$

and half-life by

$$t_{1/2} = \frac{2^{n-1} - 1}{(n-1)k[A]_o^{n-1}} \tag{8.30}$$

8.7 Dependence of velocity on temperature

It is the common experience of all that chemical reactions take place more rapidly at high temperatures. For example, oil is oxidized more quickly during frying than when stored at room temperature. Similarly, meat proteins in aqueous solutions hydrolyze more quickly when they are heated, such as during cooking. On the qualitative side, this happens because the collisions of molecules that lead to stable products increase with the increased motility of the molecules that the raised temperature brings. However, how can the dependence of reaction rate on temperature be described mathematically?

The dependence of the rate constant k of a reaction to temperature T can be represented by the *Arrhenius equation*:

$$k = Ae^{-Ea/RT} \tag{8.31}$$

Here, E_a is the *activation energy* of the reaction in question (a form of internal energy), while R is the universal gas constant. The coefficient A is called the *frequency factor* (also known as the Arrhenius factor or pre-exponential factor) and can be calculated from the linear representation of $1/T - \ln k$ from the logarithm of Equation (8.31):

$$\ln k = \ln A - \frac{E_a}{R} \frac{1}{T} \tag{8.32}$$

E_a is defined by the above equation:

$$E_a = RT^2 \frac{d \ln k}{dT} \tag{8.33}$$

If we integrate Equation (8.32) between two temperatures T_1 and T_2 we obtain:

$$\frac{d(\ln k)}{dT} = \frac{E_a}{RT^2} \Rightarrow \int_{\ln k_1}^{\ln k_2} d(\ln k) = \int_{T_1}^{T_2} \frac{E_a}{RT^2} dT \Rightarrow \ln \frac{k_2}{k_1} = \frac{E_a}{R} \left(\frac{T_2 - T_1}{T_2 T_1} \right) \quad (8.34)$$

This equation is very useful as it can easily calculate the velocity constant of a reaction when the temperature changes by ΔT. Care must be taken, however, because E_a can change with temperature, the values T_1 and T_2 have to be close to each other. Many cases also exist, especially when very low E_a values are involved where k is highly dependent on T. In such cases, more complex models than the Arrhenius equation are used.

8.8   Catalysis

Catalysis is the phenomenon whereby particular substances accelerate particular chemical reactions without being themselves included amongst the reactants or products. In general catalysts do not participate stoichiometrically in reactions since they are unbound after the end of the transformation of the reactants to products. Thus small amounts of catalyst are usually able to contribute to the reaction between many isolated molecules.

Consider a reaction A + B → C + D, which we calculate using Hess' law (see Chapter 2) to be energetically favorable; that is to say the reaction enthalpy ΔH_{re} is negative. In Figure 8.4, reactant A is represented by a white sphere joined to a black sphere and reactant B with a grey sphere.

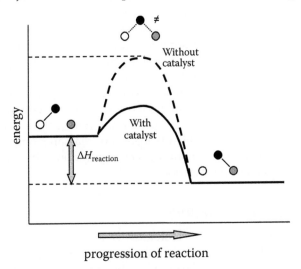

progression of reaction

Figure 8.4 Graph that describes the action of a typical catalyst. Note that the energy that is required to form the intermediate activated complex is reduced in the presence of the catalyst.

The two molecules are temporarily joined in a unified *activated complex* that is indicated by the exponent "≠". Rearrangement of bonds takes place within the activated complex: Product C is created from the connection of the black to the grey sphere, while product D is the left-over white sphere. The reaction essentially consists of the loosening of one bond and the formation of a new one.

For this to take place, two prerequisites must be met.

- There must be an overall enthalpic gain from the freeing of the old bonds and the formation of the new bonds. This is assured by the negative value of ΔH_{re}. In Figure 8.4 this prerequisite is represented by the fact that the energetic level of the products is lower than that of the reactants.
- In order to react, the two molecules must first approach each other. This requires that they concentrate at a very small part of the volume they used to occupy, i.e., increase the entropy. The energy equivalent to $T\Delta S$ must therefore be provided to the system in order for the reactant molecules to approach one another. This energy is obviously liberated ("returned") straight after the reaction, and is presented in Figure 8.4 in the form of an energetic maximum ("energetic peak") between reactants and products. An important enthalpic component can also be involved in this process.

The role of the catalyst is to reduce the height of this energetic maximum. Catalysts do not change the thermodynamic equilibrium between reactant and product, i.e., the ΔH_{re} of a reaction remains constant. Catalysts are usually substances that adsorb the reactants by means of temporary interactions with them. The enthalpic gain from such an adsorption compensates for the decrease to the entropic contribution $T\Delta S$ required for the approach of the reactant molecules to one another.

As the adsorption of large amounts of reactant requires a large surface, catalysts are usually found in colloidal dispersion or have a microporous/nanoporous structure so as to maximize the free surface per unit mass. From a kinetic perspective the adsorption of reactants onto a particular and limited space increases their local concentration, which dramatically increases the speed of the reaction.

8.9   Biocatalysts: Enzymes

Enzymes are the catalysts that biological systems use to carry out specialized reactions. They are so important to life and its functions that it is not hyperbole to maintain that "health" is the state in which an organism's enzymes are working harmoniously. Enzymic reactions continue (albeit not harmoniously) *post mortem*, and their management is one of the most important subjects in food science and technology.

Enzymes are proteins folded in such a way that a particular section of the tertiary or quaternary structure forms the *active site*, the point at which the reactants are adsorbed (*substrates* in the language of enzymology) and where the catalytic activity takes place.

As enzymes are proteins, they are particularly sensitive to changes in temperature, ionic strength, and pH. Consequently their activity is likewise sensitive to these parameters, and because of this enzymes are usually characterized by optimal ranges of pH and ionic strength for activity, while thermal denaturation inhibits their activity (as it destroys their tertiary/quaternary structure, and consequently the active site). In many cases complementary substances are necessary for the expression of enzyme activity. These substances are called *co-enzymes*. Other than their proteinaceous composition, the distinctive properties of enzymes in comparison with common catalysts are their total specialization (ability to produce very particular products from very particular substrates) and their great sensitivity to external conditions (pH, ionic strength, temperature).

8.10   The kinetics of enzymic reactions

At the beginning of the twentieth century, Michaelis, Menten, Briggs, and Haldane published a series of treatises on the kinetics of enzymic reactions. These formed the basis for the subsequent development of enzymology.[*] The aforementioned researchers reasoned that an enzyme E of concentration [E] that acts on a substrate S of concentration [S] will react reversibly with the latter, with reaction constants for the reactions $E + S \rightarrow ES$ and $ES \rightarrow E + S$ being k_1 and k_{-1}, respectively. Continuing the reaction, the activated complex dissociates into the enzyme and the product P ($ES \rightarrow P$), the reaction constant for which is k_2.

The mass balance requires that the rate (velocity) of formation of ES (which corresponds to k_1) is equal to the rate of its destruction, which occurs with the breakdown of ES to $E + S$ (constant k_{-1}) and its breakdown to $E + P$ (constant k_2). The equation of the velocities of formation and breakdown of the activated complex gives

$$k_1[E][S] = k_{-1}[ES] + k_2[ES] \Leftrightarrow k_1[E][S] - k_{-1}[ES] - k_2[ES] = 0 \quad (8.35)$$

---

[*] Beyond its obvious usefulness as the basis of enzymology, the theorem of enzymic reaction kinetics formulated by Michaelis, Menten, Briggs, and Haldane constitutes a standard of elegance and simplicity in mathematical formulation and management of physicochemical concepts. For the latter reason at least the author of the present work believes it should be taught to all students.

The overall concentration of the enzyme $[E]_o$ is equal to the concentration of the free enzyme $[E]$ plus the concentration of the enzyme in the activated complex $[ES]$:

$$[E]_o = [E] + [ES] \tag{8.36}$$

Substituting $[E]$ from Equation (8.36) into Equation (8.35), we obtain

$$k_1 \left([E]_o - [ES] \right) [S] - (k_{-1} + k_2)[ES] = 0 \tag{8.37}$$

Solving for the complex concentration, we obtain

$$[ES] = \frac{k_1 [E]_o [S]}{k_{-1} + k_2 + k_1 [S]} \tag{8.38}$$

Assuming a first-order reaction ES → P, by multiplying by k_2 we obtain the velocity V of the formation of the final product P, that is to say, the velocity with which the enzymic reaction forms products:

$$V = k_2 [ES] = \frac{k_2 k_1 [E]_o [S]}{k_{-1} + k_2 + k_1 [S]} = \frac{k_2 [E]_o [S]}{\dfrac{k_{-1} + k_2}{k_1} + [S]} \equiv \frac{k_2 [E]_o [S]}{K_M + [S]} \tag{8.39}$$

The quantity $K_M = (k_{-1} + k_2)/k_1$ is called the Michaelis constant and, as the ratio of the constants of the reactions that dissociate the activated complex to that of the reaction that forms it, it is a measure of the activity of the enzyme.[*] The final form

$$V = \frac{k_2 [E]_o [S]}{K_M + [S]} \tag{8.40}$$

is called the Michaelis–Menten equation and is sufficient to describe simple enzymic reactions. Complex reactions require the use of more complicated models.

Figure 8.5 describes a typical velocity–substrate concentration curve as given by Equation (8.40). From this it is apparent that the rate of increase

---

[*] In a more formal treatment the Michaelis constant K_m is properly defined as the substrate concentration when the velocity of the enzymic reaction has reduced to half its maximum value.

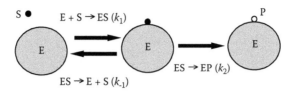

Figure 8.5 Basic concept of the Michaelis–Menten approach.

(the first derivative) of velocity declines with the substrate concentration and tends asymptotically toward a maximum value V_{max}.

If the substrate concentration [S] is very small ($K_M \gg$ [S]) it can be omitted from the denominator:

$$V = \frac{k_2}{K_M}[E]_o[S] \tag{8.41}$$

The maximal enzymic velocity V_{max} can be defined as

$$V_{max} = k_2[ES] \tag{8.42}$$

when all of the enzyme is in the form of ES.

8.10.1   Lineweaver–Burk and Eadie–Hofstee graphs

The Michaelis–Menten equation can be used to calculate the individual parameters of different enzymic reactions. Nonlinear line fitting methods are used for this purpose today. Despite this, the Michaelis–Menten equation can give linear plots which can quickly and easily provide solutions for values of K_M and V_{max}.

The simplest linear solution of the Michaelis–Menten equation is that of Lineweaver and Burke. According to this we invert Equation (8.40), obtaining the expression:

$$\frac{1}{V} = \frac{K_M + [S]}{V_{max}[S]} = \frac{1}{V_{max}} + \frac{K_M}{V_{max}}\frac{1}{[S]} \tag{8.43}$$

Equation (8.43) is called the *Lineweaver–Burke equation* and is a straight line when plotted with 1/[S] and $1/v_0$ on the x and y axes, respectively. Such a graph can be used to calculate K_M and V_{max}, as presented in Figure 8.6.

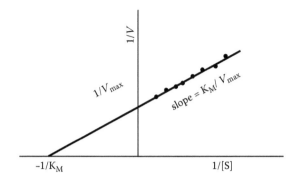

Figure 8.6 Graph and parameters according to Lineweaver–Burke derived from experimental data.

A graph of this type is plotted by measuring υ for a series of different substrate concentrations [S]. The results of the measurements are then inverted and plotted as points on a Lineweaver–Burke plot.[*]

The Lineweaver–Burke equation may introduce significant uncertainty (and consequently error) into the corresponding graph, since it inverts the measured values. For this reason many alternative linear solutions to the Michaelis–Menten equation have been developed. One of the most important alternative approaches is that of Eadie and Hofstee. This is produced as follows. Inverting the Michaelis–Menten equation and multiplying by V_{max} we obtain:

$$\frac{V_{max}}{V} = \frac{V_{max}\left(K_M + [S]\right)}{V_{max}[S]} = \frac{K_M + [S]}{[S]} \Rightarrow$$

$$V = -K_M \frac{V}{[S]} + V_{max} \Rightarrow \frac{V}{[S]} = \frac{V_{max}}{K_M} - \frac{V}{K_M}$$

(8.44)

The final phrase is called the *Eadie–Hofstee equation* and is a straight line with $V/[S]$ and V on the x and y axes, respectively. The graph of these two quantities can be used to calculate K_M and V_{max}, as shown in Figure 8.7.

The processing of the data for the Eadie–Hofstee equation avoids the inversion of υ and [S], resulting in fewer errors in the results. The

---

[*] Lineweaver–Burke plots are, because of their simplicity, suitable for an initial study of the existence of inhibitors in enzymic reactions. This subject, although very interesting, departs from the aims of the current work and for this reason is not examined.

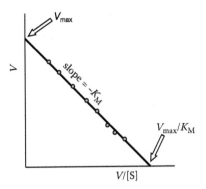

Figure 8.7 Graph and parameters according to Eadie–Hofstee derived from experimental data.

disadvantage of the method is that, since υ is present on both axes, the errors from its measurement are observed as much on the x values as on the y values.

EXERCISES

8.1 The solution of a photosensitive substance was found to have the following concentrations of C after preparation and exposure to radiation:

t (min)	0	10	50	100	150	200
C (mol L^{-1})	1.000	0.974	0.825	0.689	0.543	0.423

Calculate the order of the photodegradation reaction and the half-life.

Solution: Plot the data on graphs derived from Equations (8.11), (8.14), and (8.26) [see the corresponding figures] and see which is closest to a straight line. Calculate the half-life from the equation of the appropriate reaction order.

8.2 An acidic solution of a sugar was found experimentally to hydrolyze to 53% of its initial concentration after 60 min, following what appears to be a first-order reaction. Calculate the additional time required for the hydrolysis to proceed to 75% of the initial concentration and the extent of hydrolysis after 2 hours.

Solution: Calculate the $t_{53\%}$ considering [A] = 0.53[A]$_0$ in Equation (8.14) and from this extract the constant k. Solve then for 75% using Equation (8.14) with k known. Consider that [A]$_0$ = 100.

8.3 Reverse-engineering: Build a table of arbitrary values that obey the equations of zero-, first-, and second-order reaction kinetics (Equations (8.11), (8.14), and (8.26)), one set of values per class. Work backward in order to calculate the reaction constant k, the half-life, the activation energy, and the frequency factor.

8.4 The experimental data below were taken from the breakdown of a food additive:

T (°C)	0	30	45	65
k (min^{-1})	4.7×10^{-5}	2.5×10^{-3}	2.9×10^{-2}	3.0×10^{-1}

Calculate the activation energy and the frequency factor.

Solution: Plot a graph of $1/T$ versus $\ln k$. Approximate the parameters by fitting Equation (8.32) to the plot.

8.5 An enzyme-catalyzed reaction at 37°C and pH 6.7 was monitored photometrically. The following substrate concentration [S] and initial velocity V (μL O_2 min^{-1}) data were taken.

[S]	M/5	M/10	M/20	M/40	M/60	M/80	M/100	M/250	M/500	M/750	M/1000
V	19.1	18.1	16.8	14.4	12.5	11.2	9.9	6.9	3.2	2.6	1.6

Calculate the Michaelis constant under the conditions in question.

Solution: Construct a Lineweaver–Burke or Eadie–Hofstee plot. It is a good idea to do both and comment on the differences.

8.6 Students measured the data for an enzymic hydrolysis at 25°C and at pH 7.5 for a constant enzyme concentration. They show the initial reaction velocity V_0 (mol substrate 10^8 L^{-1} s^{-1}) as measured photometrically for different values of initial concentration of substrate [S] (mol x 10^3 L^{-1}):

V	30.3	14.5	8.41	5.05	2.47	1.41	0.39
[S]	19.7	17.2	14.9	11.5	7.8	4.8	1.7

Calculate the Michaelis constant of the enzyme and the maximum velocity under these conditions.

Solution: Construct a Lineweaver–Burke or Eadie–Hofstee plot. It is a good idea to do both and comment on the differences.

Bibliography

Adamson, A.W., and Gast, A.P. *Physical Chemistry of Surfaces* (1997). John Wiley & Sons Inc., New York.

Alberty, R.A. *Physical Chemistry 7th ed.* (1987). John Wiley & Sons, Inc, NY.

Atkins P., and de Paula J. *Physical Chemistry 7th ed.* (2002). WH Freeman & Co, NY.

Avery, H.E., and Shaw, D.J. *Basic Physical Chemistry Calculations, 2nd ed.* (1980). Butterworths, London.

Bansil, R., and Turner, B.S. Mucin structure, aggregation, physiological functions and biomedical applications (2006). *Current Opinion in Colloid and Interface Science*, (11) 164–170.

Belitz, H.-D., Grosch, W., and Schieberle, P. *Food Chemistry, 4th ed.* (2009). Springer-Verlag, Berlin.

Belton, P. (Ed.) *The Chemical Physics of Food* (2007). Blackwell Publishing Ltd., Oxford.

Birdi, K.S. (Ed.). *Handbook of Surface and Colloid Chemistry, 3rd ed.* (2009). CRC Press, Boca Raton, FL.

Blahovec, J., and Yanniotis, S. GAB generalized equation for sorption phenomena (2008). *Food Bioprocess Technology*, (1) 82–90.

Borwankar, R.P., and Case, S.E. Rheology of emulsions, foams and gels (1997). *Current Opinion in Colloid and Interface Science*, (128) 169.

Borwankar, R., and Shoemaker, C.F. (Eds.). Rheology of foods (Reproduced, from *J. Food Eng.*, Vol. 16, Issue 1,2, 168 pp.) (1992). Elsevier Applied Science, London.

Budin, I., and Szostak, J.W. Physical effects underlying the transition from primitive to modern cell membranes (2011). *Proceedings of the National Academy of Sciences of the United States of America*, (108) 5249–5254.

Cosgrove, T. (Ed.). *Colloid Science. Principles, Methods and Applications* (2005). Blackwell Publishing, Oxford.

Coultate, T.P. *Food: The Chemistry of Its Components, 2nd ed.* (1999). Royal Society of Chemistry, Cambridge.

Dalgleish, D.G., and Corredig, M. The structure of the casein micelle and its changes during processing (2012). *Annual Review of Food Science and Technology*, (3) 449–467.

De Kruif, C.G., Huppertz, T., Urban, V.S., and Petukhov, A.V. Casein micelles and their internal structure (2012). *Advances in Colloid and Interface Science*, (171–172) 36–52.

DeMan, J.M. *Principles of Food Chemistry, 3rd ed.* (1999). Aspen Publishers, Inc., Gaithersburg, MD.

Dickinson, E. *An Introduction to Food Colloids* (1994). Oxford University Press, Oxford.

Dickinson E., and Stainsby, G. *Colloids in Food* (1982). Applied Science Publishers, Barking-Essex, UK.

Dollimore, D., Spooner, P., and Turner, A. The BET method of analysis of gas adsorption data and its relevance to the calculation of surface areas (1976). *Surface Technology*, (4) 121–160.

Evans, D.F. and Wennerström, H. (1994). *The Colloidal Domain: Where Physics, Chemistry, Biology and Technology Meet*. VCH Publishers, Inc., New York.

Fellows, P.J. *Food Processing Technology. Principles and Practice* (1988). Ellis Horwood, Hemel Hempstead.

Figura, L.O., and Texeira, A.A. *Food Physics-Physical Properties-Measurement and Applications* (2007). Springer-Verlag, Berlin.

Flory, P.J. *Principles of Polymer Chemistry* (1953). Cornell University Press, Ithaca, NY.

Fox, P.F., and McSweenet, P.L.H. *Dairy Chemistry and Biochemistry* (1998). Blackie Academic & Professional, London.

Friberg, S.E., Larsson, K., and Sjöblom, J. (Eds.) *Food Emulsions, 4th ed.* (2004). Marcel Dekker Inc., New York.

Frost, A.A. and Pearson, R.G. *Kinetics and Mechanism 2nd ed.* (1961). John Wiley & Sons, Inc., New York and London.

Garti, N., and Sato, K. (Eds.). *Crystallization Processes in Fats and Lipid Systems* (2001). Marcel Dekker, New York.

Griffiths, P.J.F., and Thomas, J.R.D. *Calculations in Advanced Physical Chemistry* (1962). Edward Arnold (Publishers) Ltd., London.

Gruenwedel, D.E., and Whitaker, J.R. (Eds.). *Food Analysis: Principles and Techniques, Vol. 1: Physical Characterization* (1984). Marcel Dekker Inc., New York.

Israelachvili, J.N. *Intermolecular and Surface Forces. With Applications to Colloidal and Biological Systems (1985)*. Academic Press, London.

Holmberg, K., Jönsson, B., Kronberg, B., and Lindmann, B. *Surfactants and Polymers in Aqueous Solutions (1998)*. John Wiley & Sons Ltd., Chichester.

Karabinas, P.M., Kokkinidis, G.I., Anastopoulos-Tzamalis, A.G., and Ritzoulis, G.Ch. *Notes on the Physical Chemistry of the States of Matter and Thermodynamics.* Publishing Department, AUTh, Thessaloniki (in Greek).

Lapasin, P., and Pricl, S. *Rheology of Industrial Polysaccharides: Theory and Applications* (1995). Blackie Academic and Professional, Glasgow.

McClements, D.J. *Food Emulsions. Principles, Practice and Techniques, 2nd ed.* (2004). CRC Press, Boca Raton, FL.

McClements, D.J. (Ed). *Understanding and Controlling the Microstructure of Complex Foods* (2007). CRC Press, Boca Raton, FL.

Meierhenrich, U.J., Filippi, J.-J., Meinert, C.,Vierling, P., and Dworkin, J.P. On the origin of primitive cells: From nutrient intake to elongation of encapsulated nucleotides (2010). *Angewandte Chemie—International Edition*, (49) 3738–3750.

Misailidis, N.I. *Methods for Solving Physical Chemistry Exercises* (1994). Simonis—Chadjipantou Publications Ltd., Thessaloniki (in Greek).

Mittal, K.L., and Shah, D.O. (Eds.). *Adsorption and Aggregation of Surfactants in Solution* (2002). Marcel Dekker Inc., New York.

Murrell, J.N., Kettle, S.F.A., and Tedder, J.M. *The Chemical Bond* (1985). John Wiley & Sons Ltd., Chichester.

Myers D. *Surfactant Science and Technology, 3rd ed.* (2006). John Wiley & Sons, Hoboken, NJ.

Nakai, S., and Li-Chan, E. *Hydrophobic Interactions in Food Systems* (1988). CRC Press, Boca Raton, FL.

Panayiotou, C. *Interface Phenomena and Colloidal Systems, 2nd ed.* (1998). Zitis Publications, Thessaloniki (in Greek).

Panayiotou C. *Science and Technology of Polymers 2nd ed.* (2000). Pegasos 2000 Publications, Thessaloniki.

Price, N.C., and Dwek, R.A. *Principles and Problems in Physical Chemistry for Biochemists, 2nd ed.* (1989). Oxford Science Publications, Oxford.

Rao, M.A. *Rheology of Fluid and Semisolid Foods: Principles and Applications* (1999). Aspen Publishers Inc., Gaithersburg, MD.

Sahin, S., and Sumnu, S.G. *Physical Properties of Foods* (2006). Springer Science+Business Media, New York.

Sheludko, A. *Colloid Chemistry* (1966). Elsevier Publishing Company, Amsterdam.

Sikorsly, Z. (Ed.). *Chemical and Functional Properties of Food Components, 2nd ed.* (2002). CRC Press, Boca Raton, FL.

Silbery, R.J., Alberty, R.A., and Bawendi, M.G. *Physical Chemistry, 4th ed.* (2005). Wiley, New York.

Smith E. Brian. *Basic Chemical Thermodynamics, 2nd ed.* (1977). Oxford University Press, Oxford.

Steffe, J.F. *Rheological Methods in Food Process Engineering, 2nd ed.* (1996), Freeman Press, East Lansing, MI.

Sun, S.F. *Physical Chemistry of Macromolecules. Basic Principles and Issues* (1994). John Wiley & Sons, Inc., New York.

Thomson, A., Boland, M., and Singh, H. (Eds.). *Milk Proteins from Expression to Food* (2009). Elsevier, San Diego, CA.

Tsujiii, K. *Surface Activity* (1998). Academic Press, San Diego, CA.

Walstra, P. *Physical Chemistry of Foods* (2003). Marcel Dekker, New York.

Whitehurst, R.J. (Ed.). *Emulsifiers in Food Technology* (2004). Blackwell Publishing, Oxford.

Index

A

Abscissa, 88
Absolute zero, 3
Acid solutions, 51, 53–57, 54
Activation energy, 182
Activity, 53
  and ionic strength, 52–53
Acute angle, with high miscibility, 81
Adiabatic boundaries, 1
Adiabatic conditions, 16
Adsorption, 85
  BET isotherm, 87
  calculating using Gibbs isotherm, 84
  casein, 121–122
  conditions for proteins, 124–125
  Freundlich's equation, 87
  GAB isotherm, 89, 90
  Gibbs isotherm, 90
  Henry's equation, 86
  at interfaces, 122–123
  isotherms, 85–90
  Langmuir's isotherm, 86, 87
  protein chains onto hydrophobic
    surface, 123
  and protein denaturation, 120
  pure heat of, 87
  spherical proteins, 123–124
  thermodynamic basis, 85
  thermodynamic incentives for
    surfactants, 91
Affinity forces, 81
Agglomeration, 150
Aggregation
  destabilization by, 150–152
  mechanisms, 151
Aggregation number, 95
Alcoholic beverage production, 40
Alkaline solutions, 54

Alpha-carbon atom, 51
Alpha-form, 28
  polymorphs, 29
Amine group, 51
Amino acids, 51
  hydrophobic character, 52
  less polar, 52
  negatively charged ions, 114
  physicochemical behavior, 52
Amphiphilic molecules, xi
Arrhenius equation, 182
Arrhenius factor, 182
Atom-scale interactions, 131
  depletion flocculation, 140
  DLVO theory, 135–137
  electrostatic interactions, 134–135
  excluded volume forces, 138–141
  hydrogen bonds, 133–134
  solvation interactions, 137–138
  stereochemical interactions, 138–141
  van der Waals forces, 131–133
Attraction factor, molecular-level, 134
Attractive forces, 159
Attractive potentials, in colloidal systems,
  132
Autoclaves, 25
Available space, 4
Avogadro's constant, 19
Avogadro's number, 20, 50, 89
Azeotropic maximum, 40, 52
Azeotropic minimum, 40
Azeotropic mixtures, 41
  vapor pressure plots, 40

B

Basic solutions, 51, 53–57
Beer, froth mechanisms in, 145, 146, 148

BET isotherm, 87, 88, 89
Beta-form, 28, 29
Bi-continuous states, 106
Bimodal lines, dependence on plot shape,
    71
Binary liquid mixtures, 38, 40, 130
    phase diagram, 73
    phase separation in, 72–73
    vapor pressure change, 39
Binding energy, 133
Bingham plastics, 163
Binodal lines, 72
Biocatalysts, 184–185
Biological fluids, protein regulation in, 119
Biological systems, macromolecules in,
    59–60
Blanching, 121
Body volume, changes with tangential
    stress, 160
Boiling point, 23, 24
    elevation, 47–48
    water *versus* sulfur dioxide, 133–134
Boltzmann constant, 19, 20
Bond formation, 132
    enthalpy and, 20
    reduction of enthalpy through, 116
Bottles, 25
Boundary, 1
    adiabatic, 1
    in liquid systems, 77
    semipermeable, 1
Boyle–Mariotte equation of state, 9
Branch formation, 105
Bread products, 125
    bubble phenomena, 145
    covalent disulfide bonds, 113
Bridging, 151, 152
Brownian motion, 127, 150
    in colloidal systems, 128
Bubble coalescence, protecting against, 145
Bubbles, 150
    in foams, 130
Buffer solutions, 53–57, 56
Bulk modulus, 160
Butter, as emulsion, 128

C

Canning, 24
Capillary action, 82
Capillary effects, 79, 81–82
Carbonated soft drinks, bubble formation,
    148

Casein
    flocculation, 116
    isoelectric point, 115
    structure, self-assembly, adsorption,
        121–122
Catalysis, 183–184
Catalysts, action of typical, 183
Cavitation, 143
Charged polysaccharides, 57
Cheese-making, 122
    bubble phenomena, 145
Chemical bonds, 22
Chemical equilibrium, 41–44
    in solutions, 44–46
Chemical kinetics, xi, 173
    biocatalysts, 184–185
    catalysis, 183–184
    enzymes, 184–185
    enzymic reactions, 185–189
    exercises, 189–190
    first-order reactions, 177–182
    higher-order reactions, 180–182
    kinetic energy requirements, 173–174
    reaction laws, 174–176
    second-order reactions, 180–182
    velocity, 174
    velocity dependence on temperature,
        182–183
    zero-order reactions, 176–177
Chemical potential, 31–32, 46–47, 81
Chemical reactions, acceleration through
    adsorption, 85
Chemical thermodynamics, 13
    application of phase transitions, 27–31
    beyond temperature, 13–15
    chemical potential, 31–32
    crystallization, 27
    entropy, 17–21
    exercises, 32–33
    fat melting, solidification, and
        crystallization, 27–31
    phase transitions, 21–27
    thermochemistry, 16–17
Chemistry
    compressibility, 4–6
    deviations from ideal behavior, 4–10
    exercises, 10–11
    gas molecule schematics, 7
    isobaric transformation, 4
    isothermic transformation, 2
    physical basis, 1
    real gas isotherms, 8
    solutions to van der Waals equation, 8

temperature, 2–4
thermodynamic systems, 1–2
triple solutions to van der Waals
equation, 8
van der Waals equation, 6–9
virial equation, 9–10
Chlorine-substituted benzenes, polarity, 51
Chocolate, 30–31
Chymosin, 122
Clapeyron diagram, 2
Clapeyron equation, 26
Clasius–Clapeyron equation, 26, 27
Closed systems, 1
Coalescence
destabilization by, 152
prevention strategies, 154, 155
Coatings, 93
Cocoa butter, 30–31
Cohesive forces, 81
Collagen, 119
Colloidal calcium phosphate, 116
Colloidal dispersions, 91, 93
rheology of, 105
Colloidal systems, xi, 127–128, 165
Brownian motion in, 127-128
DLVO theory, 135–137
electrostatic stabilization, 135–137
emulsions and foams nomenclature,
128–130
examples and physical states, 128
large surface areas in, 127
light scattering in, 146–147
Colloids, with polydisperse particle sizes,
149
Complex modulus, 170
Complex rheological behaviors, 165
in emulsions and foams, 165–167
Component separation, through
adsorption, 85
Compressibility, 4–6
Compressibility coefficient, 5
and pressure, 5
Compressibility factor, 6, 9
at high pressures, 6
Compression, volume of gas under, 7
Concentrated polymer solutions, 69–70
Concentrated regime, 70
Condensation, 9, 18
micelles and, 95
Condensation point, 22
Conductivity, 97
sudden change in phase inversion, 153
Cone shapes, 100, 104, 107

Configuration entropy, 139
Conformational entropy, 61–65, 72, 117
increase with protein folding, 114
Conjugate base, 55
Contact angle, 81
in detergents, 144
droplet on surface, 81
Continuous aqueous phase, 90, 112
Continuous phases, 106, 137
in emulsions, 128
of emulsions and foams, 127
viscosity, 149
Continuous triglyceride phase, 90
Coordination number, 62
Correlation length, 69, 70
Cosmetics science, xi
Counter-enthalmic principle, 18
Counter-ions, 134, 135
Covalent disulfide bonds, in bread
products, 113
Covolume, 6
Cream, 165
Creaming, 148–150
of large isolated flocs, 151
prevention strategies, 155
Creep compliance, 168
Creep method, 168
Critical micelle concentration (cmc), 96–98,
153
spontaneous micelle creation above, 97
Critical packing parameter, 103
Critical point, 24, 25
Critical pressure, 24
Critical temperature, 7, 24
Crystal lattice, 27, 29
Crystallization, 27
of fats, 27–31
Crystals, 27
Cubic phase, 106, 108
hydrophobic heads arrangement, 107
Curved surfaces, work required to form,
142
Cylindrical micelles, 100, 104, 107–108

D

Dairy products
destabilization by polysaccharides, 48
milk scalding, 117
protein precipitation and flocculation
in, 115
Deborah number, 157
Decontaminants, 93

Deformation, 157, 160
    due to shear stress, 167
    irreversible, 161
    rate of, 161
    reversibility in solids, 169
    in solids, 169
    time-dependent, 168
Degrees of freedom, 18
    deprivation, 27
    increased with thermal motion, 19
    maximization by fats, 27
Denaturation. *See* Proteins
Dense emulsions, 162
    volume fraction and viscosity, 167
Depletion flocculation, ix, 140, 141, 151
Depletion interactions, 139, 141
Depth, and velocity in Newtonian flow, 162
Desorption, 124
Destabilization
    aggregation mechanisms, 150–152
    by coalescence, 152
    by creaming, 148–150
    by disproportionation, 153–154
    of emulsions and foams, 147–154
    five mechanisms of, 148
    by flocculation, 150–152
    by gravitational separation, 148–150
    by Ostwald ripening, 153–154
    by phase inversion, 153
Detergents, 139, 165
    as archetypal emulsifiers, 144–145
    misperceptions, 144
Diamonds, kinetic stability, 173–174
Dielectric constant, 50
Diffusion, in small droplets, 148
Dilatant thickening fluids, 162–163
Dilute solutions, 69
Dipole moment, 50–52
Dipoles, 132, 134
Direct interaction, 73
Disorder, 21
Dispersed phase
    becoming continuous, 153
    colloidal system examples, 128
    in emulsions, 128
    emulsions and foams, 127
Disproportionation, destabilization by, 153–154
Dissociation, of acids, 56, 57
Dissociation constant, 53, 54
Dissociation energy, 133
Distillation, 39
    sequential, 40

DLVO potential, 136
    and distance between approaching surfaces, 136
DLVO theory, 135–137
Donnan effect, 68–69
Double bonds
    chain curvature under, 30
    straightening issues, 29
Dough, 163
Drainage, 146
    in foams, 130
Droplet formation, in emulsions, 143
Droplet on surface
    contribution of surface tension to shape, 81
    schematic representation, 81
Droplet shapes, spherical, 129
Droplet size
    in emulsions, 129
    factors determining, 143
Droplets
    clustering, 150
    collision between, 166
    deformability in shear flow, 149
    depletion flocculation, 140
    effects of flocculated, 151
    in emulsions, 129
    factors in light scattering, 146
    high concentration effects, 150
    ion distribution on surfaces, 135
    osmotic pressure, 139
    production requirements in emulsions, 142
    structure-spanning network, 150
Dyes, 93

E

Eadie–Hofstee equation, 188
Eadie–Hofstee graph, 187–189
Egg and lemon sauce preparation, pH control, 116
Einstein's equation, viscosity, 167
Einstein's law, 14
    and rheology of colloidal dispersion, 105
Elastic behavior, 157, 159–161
Elastic modulus, 159
Elasticity, 161, 169
Electrical double layer, 134
Electrolytes, 53
Electron cloud, center of density, 132
Electrostatic effects, 53
Electrostatic interactions, 134–135

Electrostatic potentials, 136
  in colloids, 135
Electrostatic stabilization, of colloids,
    135–137
Elongated cylinders, 108
Empty space, 4
Emulsification, 145–144
  in detergents, 144–145
  as energy-consuming process, 142
  role of mechanical energy, 145
Emulsifiers, 93, 129
  detergents as archetypal, 144–145
  hydrophobic surfactants, 97
Emulsions, xi, 127
  atom-scale interactions, 131–141
  change in viscosity with flocculation, 166
  colloidal systems and, 127–130
  defined, 128
  dense, 162
  destabilization, 147–154
  energy requirements, 173
  five destabilization mechanisms, 148
  increased viscosity, 166–167
  as inherently unstable systems, 141
  inverse micelles, 111
  light scattering from colloids, 146–147
  multiple scattering, 147
  nomenclature, 128–130
  production from immiscible liquids, 142
  rheology, 165–167
  stabilization against recoalescence, 141
  structure changes as function of
      volume fraction, 166
  thermodynamic considerations, 130–131
  trapped energy in, 141
Endothermic reactions, 16
Energetic gain, 115
Energetic transformations, 14
Energy
  as driver of spontaneous change, 19
  heat as, 13
  mass conversion to, 14
Energy gain per surface unit area, 80
Energy requirements, catalysis, 183
Enthalpic gain, 184
Enthalpic process, 61, 134, 161
  contribution to colloidal systems, 131
  role in miscibility, 130
Enthalpy, 16, 116
  and bond formation, 20
  change in, 17
  decrease in, 21
  of micelle formation, 102

  minimization, 117
  of mixing, 66–67
  as principle of reduction, 18
  protein folding and bond formation
      effects, 116
  in proteins, 123–124
  reduction with vaporization, 23
Enthalpy of combustion, 16
Enthalpy of formation, 16
Enthalpy of fusion, 23
Enthalpy of reaction, 16
Entropic component, 20, 21, 24
Entropic process, 61
Entropy, 17–21, 19, 64, 81
  change during expansion, 20
  conformational, 61–65
  and degrees of freedom, 18
  with even molecular distribution, 48
  and free movement, 22
  increase in, 22
  maximization, 116
  and melting, 23
  of micelle formation, 102
  of mixing, 61–65, 134
  reduction in, 27
  reduction through adsorption, 85
  reduction through protein folding, 116
  role in miscibility, 130
  in solutions, 58
Entropy of vaporization, 26
Environment, 1
Enzyme concentration, 186
Enzyme substrates, 185
Enzymes, 184–185
  sensitivity to temperature, ionic
      strength, pH, 185
Enzymic reactions
  calculating parameters of, 187
  Eadie–Hofstee graph, 187–189
  kinetics, 185–187
  Lineweaver–Burke graph, 187–189
Equilibrium, 26
  between free and polymeric surfactant
      concentration, 101
  process, 32
  between solution an its vapor, 46
  systems in, 1
Equilibrium constant, 43, 44, 45
Equilibrium point, 32
Equilibrium vapor composition, 38
Equivalent volume, 26
Evaporation, of liquids, 37
Excluded volume forces, 5, 6, 72, 138–141

Exercises
  chemical kinetics, 189–190
  chemical thermodynamics, 32–33
  chemistry physical basis, 10–11
  solution thermodynamics, 73–75
Exothermic reactions, 16
  micelles and, 95
Extended aliphatic chains, 51
External force fields, 157

F

Fat-water interface, Pickering stabilization, 144
Fats, melting, solidifying, crystallization, 27–31
Fatty acids, 28, 59
Fermentation reactions, 115
Fibrils, 119
Final concentration profile, 134
Final state, 13, 14
First derivative, 71
First Law of Thermodynamics, 16
First-order reactions, 177–178
  course of, 178
  inversion of sucrose, 178–180
Flocculate collapse, 167
Flocculated emulsion, 165
  viscosity change, 166
Flocculation, 137
  denaturation and, 118
  depletion-type between droplets, 140
  destabilization by, 150–152
  irreversible bridging, 151
  mechanisms, 151
  mucins and, 120
  obstacles to induced deformation, 167
  of proteins, 115
  structure-spanning network created by, 150
  and variation in particle sizes, 150
Flory–Huggins approximation, 66, 72
Flory–Huggins parameter, 67
Flory–Huggins theory of polymer solutions, 60–61, 68
  conformational entropy, 61–65
  enthalpy of mixing, 66–67
  entropy of mixing, 61–65
Flow, 157–159
  perception of, 157
  in thixotropic and rheopectic materials, 164
Flow behavior index, 163

Fluid flow, 163
Fluids, 157
  viscous character, 169
Fluorimetry, 97
Foaming, 145–146
Foaming agents, 93
Foams, xi, 127
  atom-scale interactions, 131–141
  colloidal systems and, 127–130
  defined, 130
  destabilization mechanisms, 147–154
  energy requirements, 173
  foaming process, 145–156
  as inherently unstable systems, 141
  nomenclature, 128–130
  rheology, 165–167
  stabilization until expiry date, 147
  thermodynamic considerations, 130–131
  trapped energy in, 141
Fontell diagram, 110
Food, macromolecules in, 59–60
Food behavior, physicochemical terms, ix
Food cans, 25
Food science, 93, 115
  emulsions and foams, 130
  protein adsorption examples, 139
  protein denaturation, 117–118
Forces
  in colloidal systems, 131
  between molecules, 10
Form, preservation, 18
Fractal dimensionality, 151, 152
Fractional distillation, 38–41, 39
Free energy
  carbon forms, 178
  change due to osmotic pressure, 49
  change during transformation, 42
  change under constant temperature, 49
  versus chemical potential, 46
  differences between reactants and products, 43
  due to surface tension, 77, 78
  minimization of surface area-to-volume ratio and, 79
  of mixing, 71, 131
  negative for protein denaturation, 120
  protein folding and, 116
  role in emulsion destabilization, 129
Free movement
  and entropy values, 22
  loss through adsorption, 85
  maximization, 21
  in protein folding, 117

Free surface area
   minimizing via spherical shapes, 79
   reduction with protein folding, 114
Free surface energy, reducing via weak
      forces, 80
Free triglycerides, 26
Freezing point, 22
   depression of, 47–48
Frequency factor, 182
Freundlich isotherm, 87
Froth, mechanisms in beer, 145
Fusion, enthalpy of, 23

G

GAB isotherm, 89, 90
Gas bubbles, 150
Gas-liquid interface, foams at, 141
Gas molecules
   critical temperature, 7
   dispersion into liquids/gels, 145
   schematic representation, 7
Gas physics, 1–2
Gaseous phase, 7, 25
   transition to, 22
Gases
   heat capacity, 13
   viscosity of, 158
Gel formation, 109
Gelatin, 119
Gels, 145
   as colloids, 127
   rheological flow, 168
Geometric shapes
   in emulsions, 142
   surfactants, 99
Gibbs' approach, to change in property,
      83
Gibbs diagram, triangular, 112
Gibbs free energy, 21, 26, 31, 43, 130, 141
   of mixing, 67, 131
Gibbs isotherm, 84, 85, 90
Gibbs–Marangoni effect, 152
Gibbs surface, 82, 83
Gluconic acid, 116
Glutamine, 52
Glycoproteins, 120
Graphite, *versus* diamond, 173
Gravitational separation, 130, 148–150
   prevention strategies, 155
Grease, as thixotropic fluid, 164
Grid cell exchange, for solvent and solute,
      58

H

Half-life, 181
Heads, 91
Heat
   under constant volume, 15
   as form of energy, 13
   transformation into work, 13
Heat capacity, 13
Heat transfer
   in endothermic and exothermic
         reactions, 16
   negative, 16
   positive, 16
Heating, and protein denaturation, 120
Helium, bond formations, 132
Henderson-Hasselbalch equation, 56
Henry's law of adsorption, 86
Herschel–Bulkley model, 163
Hess' law, 17, 173
Heterogeneous systems, 1
High pressure, gas behavior at, 6
Higher-order reactions, 180–182
Hollow micelles, 100, 109–110
Homogeneity, scale and, 1
Homogeneous systems, 1
Homogenization, 129
Hooke's constant, 159
Hooke's law, 159–161
Hurdle technology, 121
Hydration interactions, 137
Hydrocarbon aggregations, 94
Hydrochloric acid, 41
Hydrodynamic phenomena, 149
Hydrogen bonds, 113, 115, 133–134
   in water, 77
Hydrogenated fats, solid state, 30
Hydrolysis, meat proteins, 182
Hydronium, 54
Hydrophilic interactions, and surfactants,
      90
Hydrophilic-lipophilic balance (HLB), 96–98
Hydrophobic chain length, and molecule
      solubility, 97
Hydrophobic heads, arrangement in cubic
      phase, 107
Hydrophobic interactions, 137
   between methyl groups, 27
   in proteins, 124
   and surfactants, 90
Hydrophobic molecules
   distancing of, 94
   preference for nonpolar environment, 94

Hydrophobic tails, 95
  surfactant with, 99, 111
  surfactants with two, 104
Hydrostatic pressure, 159
Hysteresis, 164

I

Ice cream
  bubble phenomena, 145
  as emulsion/foam, 130
  partial coalescence in, 152
  quaternary structure roles, 125
Ideal behavior, deviations from, 4–10
Ideal gas, xi, 2, 26
  expansion, 19
  to ideal solutions from, 35–38
  proportionality constant for, 3
  real gases approaching, 6
  similarity of noble gases, 5
  viscosity, 158
Ideal gas equation, 3, 4, 5
  gas molecule schematic representation,
    7
Ideal gases, 5
Ideal plastics, 163
Ideal solutions, 36
  chemical potential approach, 46–47
  from ideal gases to, 35–38
Immiscibility
  emulsions and, 142
  between phases, 128
Immiscible liquids, 80
  interface definition, 82
  property changes, 82
Incident radiation wavelength, light
    scattering and, 146
Induced dipoles, 132
Initial state, 13, 14
  same as final state, 15
Inorganic salts, 92
  concentration at interface, 85
Instability. *See also* Destabilization
  of emulsions and foams, 127
  and surface tension, 77
Instantaneous dipoles, 132
Interactive potential, 135
Interchange energy, per solvent molecule,
    67
Interface, counter-ion congregation at, 134
Interface definition, 82–83
Interface tension, 79–80
  special extended case, 80–81

Interfaces
  adsorption and self-assembly at,
    122–123
  distance and DLVO potential, 136
  positioning by surfactants, 93
  property change and distance from, 83
  protein behavior at, 122
  with surfactant molecule layer, 91
  total energy during expansion/
    contraction, 80
  variable energetic properties, 82
Interfacial film-rupture, 152
Interfacial tension, 129
Intermediate states, 14
Intermolecular attraction coefficient, 6
Intermolecular interactions, 35
Internal energy, 14
  change in, 14
  ideal gas, 19
Intrinsic viscosity, 60
Inverse cubic phase, 106
Inverse micelles, 111, 112
Inverse structures, 110–112
Inversion, of sucrose, 178–180
Inverted cones, 104
Inverted truncated cones, 104
Ion distribution, on droplet surface, 135
Ionic strength, 52–53, 118
  enzyme sensitivity to, 185
Irreversible bridging flocculation, 151
Irreversible reactions, 29, 41, 42
  protein absorption at interfaces, 125
  protein denaturation, 118
  protein unfolding, 117–118
Isobaric conditions, 21, 24
  enthalpy of combustion in, 16
  heat associated with, 16
Isobaric transformation, 3
  temperature and volume plot, 4
Isobaric work, 25
Isochoric change, 3
Isoelectric point, 57, 115, 116
Isolated systems, 1
Isothermic transformation, 3
  pressure and volume curves, 2
Isotherms, 2
  gradient at specific temperature, 2
  real gases, 8
Isotropic bi-continuous state, 106

J

Jars, 25

K

Kamerlingh–Onnes equation, 9
Keesom forces, 133
Kelvin scale, 3
Kinetic stabilization, 117, 147
    diamonds and, 173–174
    of emulsions and foams, 141
Kinetics, 173
    enzymic reactions, 185–187
Krafft point, 96–98, 153

L

Lactic acid, 116
Lamellae, 106
    cross-section, 100
    membranes, 108–109
Laminar flow, 142, 143
Langmuir's isotherm, 86, 87
Laplace pressure, 142
Laplace's law, 17
Latent heat of fusion, 48
Latent heat of vaporization, 23, 27
Lavoisier's law, 17
LeChatelier's principle, 56, 85
Lennard–Jones equation, 132
Levinthal paradox, 116–117
Life, origins in surfactants, 93, 110
Light scattering, 127
    in colloids, 146–147
Like dissolves like, 93
Line tension, 79
Linear deformation, 159
Linear flow, 142
Linear stress, consequences, 160
Linear viscoelasticity, 168
Lineweaver–Burke graph, 187–189
Liposomes, cross-section, 100
Liquid–liquid colloids, 128. *See also*
    Emulsions
    emulsions as, 141
Liquid phase, 7, 21, 24
    transition from gas to, 7
    triglycerides, 29
    viscosity, 158
Liquid phase composition, 38
Liquid solutions, 35
Liquid surface geometry, 81–82
Liquid systems, boundaries, 77
Liquids, viscosity of, 158
Liquids in tubes, 82

Local concentration, increase through
    adsorption, 85
London forces, 132, 138
Loops, 139
Loss modulus, 170
LSW equation, 154
LSW theory, 154
Lyophobic dispersions, 127
Lyophobic interactions, 137
Lyophobic repulsions, 124

M

Macromolecular chain, 58
Macromolecules, 59, 164
    in food, 59–60
    intrinsic viscosity, 60
    osmotic pressure of solutions, 68–69
    self-assembly, 112–113
    in solution, 57–58
Mark–Houwink equation, 60
Mass, conversion to energy, 14
Material flow, 157–159
Materials science, xi
Meat proteins, hydrolysis, 182
Melting, 22, 26
    of fats, 27–31
Melting point, 22, 23
    alpha-polymorphs, 29
    chocolate, 30–31
    triglycerides, 28
    water *versus* sulfur dioxide, 133–134
Membranes, 108–109
    3D section, 109
Memory, of elastic material, 168
Metastable states, 7
Metastable systems, 72
Methyl groups, 28
    hydrophobic interactions between, 27
Micelles, 94–96
    branch formation, 105
    casein, 121
    continual exchange of molecules, 96
    cross-section, 100
    cylindrical, 107–108
    deviations from spherical, 98–100
    geometric shapes, 99
    hexagonal phase, 106
    hollow, 109–110
    rod-shaped, 104
    as simplest form of aggregate, 94
    spherical, 107
    spontaneous creation above cmc, 97

stabilization phase, 96
standard energy of formation, 101, 102
standard enthalpy, 102
standard entropy, 102
as supermolecular structure of
    surfactants, 95
Michaelis–Menten approach, 186, 187
Microemulsions, 129
Milk
    transition to yogurt, 170, 171
    whiteness phenomena, 146
Milk proteins, 121
Miscibility
    and contact angle, 81
    and entropic/enthalmic factors, 130
Mixing
    enthalpy of, 66–67
    entropy of, 61–65
    Gibbs free energy, 67
    versus phase separation, 80
Molar fraction, 36, 38, 47, 81
    polymer in solution, 71
Molar heat capacity, 13
Molar polarity, 50
Molecular dipolar torque, 50
Molecule spreading, 19
Molecules, surfactant prerequisites, 90
Molten globules, 124
Molten polymers, 162
Momentum transfer, 159
    after shear stress, 158
Monomers, 61
    condensation to polymeric micelle, 101
    density distribution with distance, 138
Monophasic systems, 1
Monotropic reactions, 29
Monounsaturated fatty acids, 28
Motility, and number of attractive
    interactions, 5
Mucins, 119, 120
Mutual interactions
    gases without strong, 5
    increased, 18
    lack of strong, 5

N

Negatively charged ions, 114
Neutral solutions, 51, 54
Newtonian flow, 158, 161–162
    typical, 162
Newtonian fluids, 162
Noble gases, 5

Non-Newtonian flow, 162
    in emulsions and foams, 165–167
    time-dependent, 164–165
    time-independent, 162–163
Non-Newtonian fluids, 162
Nonadsorbable hydrocolloids, 141
Nonpolar molecules, 50
    polarization by permanent dipole, 133
    role in surfactants, 90
    and spherical proteins, 113
    surfactants, 93
    in water, 91
Nuclear magnetic resonance (NMR)
    spectroscopy, 97
Nucleic acids, 59

O

Obtuse angle, with low miscibility, 81
Oil-in-water emulsions, 97, 128, 129, 147
Oligosaccharides, 120
One-stage distillation, 39
Open systems, 1
Organized movement, 13
Oscillation methods, 169–170
Osmotic pressure, 48–50, 49, 129, 138, 140
    Donnan effect, 68–69
    forcing spherical droplets together, 140,
        141
    solutions of macromolecules, 68–69
Ostwald ripening, ix, 130, 142, 154
    destabilization by, 153–154
    prevention strategies, 155

P

P-versus-V diagram, 7, 8
Partial coalescence, 152
Partial pressure, 38
    and vapor pressure, 37
Partially miscible liquids, 131
Particle size
    and light scattering, 146
    variation and flocculation, 150
Peptide chains, 119
Permeable boundaries, 1
pH
    bringing to isoelectric point, 115
    change with protein denaturation, 120
    in cheese-making, 122
    in egg and lemon sauce preparation, 116
    enzyme sensitivity to, 185

protein solutions, 115
in solutions, 53–57
Phase change
chocolate example, 30–31
cmc and, 97
latent heat, 26
Phase diagrams, 24, 112
Phase inversion, destabilization by, 153
Phase inversion temperature (PIT), 153
Phase separation, 70–72, 73, 80, 94
causes in emulsions, 141
by mucins, 120
in two-solute systems, 72–73
Phase transitions, 21–27
application to fats, 27–31
Phosphorylated serine, 121
Physical chemistry, xi
versus chemical kinetics, 174
Pickering stabilization, 130, 144
in foaming processes, 145
Plastic flow, equations, 163
Point tension, 79
Polar hydroxyl groups, 113
Polar molecules, 50, 51
and elongated proteins, 113
inorganic salt interactions, 92
role in surfactants, 90
surfactants, 93
in water, 91
Polarity
application to proteins, 51–52
chlorine-substituted benzenes, 51
and dipole moment, 52–52
Polydisperse particle sizes, 149
Polyelectrolytes, 68
Polymer chain, 113
Polymer dissolution, spontaneous, 61
Polymer solutions, 59
concentrated, 69–70
Flory–Huggens theory, 60–61
molar fraction, 71
phase diagram, 73
specific viscosity, 60
Polymers
micelle formation, 101
properties, xi
in solution, 58–59
visualization, 2D grid, 59
Polymorphism
schematic, 29
in triglycerides, 28
Polyprotic acid, 56
Polysaccharides, 58, 59, 113

Polyunsaturated fatty acids, 30
Porosity, 89
Positively charged ions, 115
Pre-exponential factor, 182
Pressure, 1
and compressibility coefficient, 5
compressibility coefficient and, 5
relationship to temperature and
volume, 2
state transformation under constant, 15
Pressure cooking, 121
Property change, and distance between
interface, 83
Proportionality constant, 159
for ideal gases, 3
in tangential tension, 160
Proteins, 51, 59
adsorption conditions, 124–125
adsorption onto hydrophobic surface,
123
behavior at interfaces, 122
behavior in solutions, 114–116
blanching and denaturation, 121
denaturation, 117, 118, 120–121
desorption, 124
dislocation, 125
effects of heating, 117–118
effects of solvents, 118–119
effects on solvents, 119–120
elongated forms, 113
enzymes as special-case folded, 185
flocculation, 115
as interface-active agents, 122
loss of tertiary and secondary
structures, 124
lyophobic repulsions, 124
polar and nonpolar amino acids in, 113
polarity and, 51–52
quaternary structures, 113
self-assembly, 112–113
self-folding, 116–117
suspension in aqueous solutions, 115
thermodynamic incentive for folding,
114
total free energy gain of folding, 117
unfolding, 114
variable shapes, 114
weak intramolecular interactions, 123
Protons, 55
Pseudoplasticity, 162, 165
Pseudosolutions, 127
Pure heat of adsorption, 87

Q

Quaternary structures, 113, 122, 125
  as active site for enzymes, 185
Quicksand, 163

R

Random motility, 18
Random movement, 13, 20
Raoult's law, 36, 38, 41, 46, 52, 59
Reaction constant, 44
Reaction laws, 174–176
Reaction rate, dependence on temperature, 182
Reactions, in solutions, 44
Real gases
  isotherms, 8
  similarity to ideal gases, 6
Real solutions, activity and ionic strength, 52–53
Real volume, 5
Recoalescence
  of newly formed droplets, 143
  stabilization against, 141
Recovery, after removal of stress, 168
Reduction, principle of, 18
Reference pressure, 36
Refractive index, 146
Relaxation, 169
Rennet, 122
Reversible processes, 41
  protein folding and unfolding, 116
Reversible reactions, 16, 23, 25
Reynolds number, 143
Rheology, xi, 157
  of colloidal dispersions, 105
  complex behaviors, 165–167
  elastic behavior, 159–161
  of emulsions and foams, 165–167
  gel flows, 168
  Hooke's law, 159–161
  material flow, 157–159
  methods for determining viscoelasticity, 168–170
  momentum transfer after shear stress, 158
  mucins in, 119–120
  Newtonian flow, 161–162
  non-Newtonian flow, 162–165
  protein roles in regulating, 119
  viscoelasticity, 168
  viscosity change, flocculated emulsion, 16
  viscous behavior, 161–162
Rheopectic materials, 168
  defined, 164
  flow curves, 164
  viscosity of, 165
Rigid molecules, 57–58
Rod-shaped micelles, 104, 105, 108
Rotation, 58

S

Salt solutions
  behavior at different concentrations, 49
  lowering of cmc by, 98
Salting out, 119
Salts, destabilization of emulsions/foams by, 137
Sausages, as emulsion, 128
Scale, and homogeneity, 1
Second derivative, 72
Second Law of Thermodynamics, xi, 41
Second-order reactions, 180–182
  course of, 181
Selective permeability, 109
Self-assembly
  as anti-entropic process, 93
  casein, 121–122
  in food science, 93
  hydrocarbon aggregations, 94
  at interfaces, 122–123
  micelles, 96
  and origins of life, 110
  required forces, 96
  structures resulting from, 104–112
  thermodynamics, 100–104
Semi-dilute limit concentration, 105, 106
Semifluid flow, 163
Semipermeable boundaries, 1
Semipermeable membranes, 48, 49
Separation distance, 18
Sequential distillation, 40
Serine, 52
Shear force, 158
Shear modulus, 160
Shear stress
  and deformation, 167
  momentum transfer after, 158
Shear thickening fluids, 162–163
Shear thinning, 162
Shearing, rate of, 161
Solid body behavior, 159

Solid phase, 21, 25
    viscosity, 158
Solidification, of fats, 27–31
Solids, 5, 157
    deformation and applied stress, 169
    elasticity of, 158
    entropic factor, 22
Solubilization, 70
    and Krafft point, 98
Solution thermodynamics, 35
    acid, base, and buffer solutions, 53–57
    activity and ionic strength, 52–53
    boiling point elevation, 47–48
    chemical equilibrium, 41–44
    chemical equilibrium in solutions,
        44–46
    chemical potential approach, 46–47
    concentrated polymer solutions, 69–70
    differences from food science, 93
    dipole moment and, 50–52
    Donnan effect, 68–69
    exercises, 73–75
    Flory–Huggins theory of polymer
        solutions, 60–67
    fractional distillation, 38–41
    freezing point depression, 47–48
    ideal gases to ideal solutions, 35–38
    ideal solutions, 46–47
    intrinsic viscosity and, 60
    macromolecule solution osmotic
        pressure, 68–69
    and macromolecules in food, 59–60
    macromolecules in solution, 57–58
    osmotic pressure, 48–50
    pH and, 53–57
    phase separation, 70–73
    polarity and, 50–52
    polymers and, 58–59
    proteins and polarity, 51–52
    real solutions, 52–53
Solutions
    chemical equilibrium in, 44–46
    macromolecules in, 57–58
    properties, xi
    protein behavior in, 114–116
Solvation interactions, 137–138
Solvent viscosity, 106
Solvents
    effects of proteins on, 119–120
    effects on proteins, 118–119
Space occupation, 4–5, 18, 19
Sphere volume, 5
Spherical droplets, 129

    in emulsions, 142
    osmotic pressure on, 140, 141
Spherical micelles, 103, 107
    critical packing parameter, 103
    deviations from, 98–100
    shape stability, 108
Spherical proteins, 125
    adsorption, 123–124
    denaturation, 118
    nonpolar amino acid content, 113
Spherical shapes, surface tension and, 79
Spinodal lines, 72
    dependence, 71
Spontaneous transformations, 17, 100
Stability
    DLVO theory insights, 137
    graphite *versus* diamond, 173
    at high polymer concentrations, 70
    hydrogen bonds role in, 133
Stability regimes, 71
Standard concentrations, 44
Starch pastes, 163
State changes, 14
    under constant pressure, 15
State prediction, 112
Stereochemical interactions, 138–141
Steric interactions, 138
    colloid stabilization, 139
Stern layer, 134
Stirling approximation, 63
Stokes equation, 149
Storage modulus, 170
Stress, solid body behavior during, 159
Strong axis, dissociation, 55
Strong bonds, 35
Structure
    casein, 121–122
    effect on cmc, 97
    formation through adsorption, 85
    preservation, 18
    self-assembly forms, 104–112
    succession from micelle to lamella, 111
Sublimation, 26
Sublimation point, 24
Sucrose, inversion, 178–180
Sulfur dioxide, as cousin molecule to water,
        133–134
Supercritical state, 24, 25
Surface-active materials, 93
    adsorption and self-assembly at
        interface, 122–123
    casein structure, self-assembly,
        adsorption, 121–122

critical micelle concentration (cmc),
  96–98
cylindrical micelles, 107–108
defined, 93
deviations from spherical micelle,
  98–100
hexagonal phase, 105
hollow micelles, 109–110
hydrophilic-lipophilic balance (HLB),
  96–98
hydrophobic heads in cubic phase, 107
inverse structures, 110–112
Krafft point, 96–98
lamellae, 100, 108–109
Levinthal paradox, 116–117
liposomes, 100
location, 93
macromolecule self-assembly, 112–125
micelles, 94–96
phase diagrams, 112
protein behavior in solution, 114–116
protein denaturation, 120–121
protein effects on solvents, 119–120
protein folding, 116–117
protein heating effects, 117–118
protein self-assembly, 112–125
protein shapes, 114
self-assembly thermodynamics, 100–104
self-organized structures in cross-
  section, 100
solvent effects on proteins, 118–119
spherical micelles, 107
spherical protein adsorption, 123–124
structures resulting from self-assembly,
  104–112
Surface activity, xi, 77, 83–85
  adsorption, 85–90
  capillary effects, 81–82
  contact angle, 81
  Gibbs' approach, 83
  interface definition, 82–83
  interface tension, 79–81
  liquid surface geometry, 81–82
  special extended case, 80–81
  surface tension, 77–79
  surfactants, 90–92
  wire frame experiment, 77, 78
Surface area
  calculating with BET isotherm, 88
  of colloidal systems, 127
  per unit volume, 142
Surface area-to-volume ratios,
  minimization and free energy, 79

Surface concentration, calculating with
  Gibbs isotherm, 90
Surface dissociation, 80
Surface tension, 77–79, 90, 97
  contribution to droplet shapes, 81
  defined, 79
  following adsorption at interface, 84
  reduction due to adsorption, 90
Surfactant concentration, and structure
  succession, 111
Surfactant molecules
  and Gibbs–Marangoni effect, 152
  micelle creation above cmc, 97
  nonpolar and polar chemical nature, 93
  and origin of life, 93
  tail length, 103
Surfactants, 90–92, 143
  concentration, 105
  cone-shaped, 111
  geometric shapes, 99
  hexagonal phase, 105
  interface representation, 91
  lamella of two-tailed, 109
  self-assembly, 94
  successive layers, 100
  technological parameters, 96
  thermodynaimc incentives for
    adsorption, 91
  thermodynamic complications, 91
  with two hydrophobic tails, 111
Surroundings, 1
System properties, changes and interface,
  82
System viscosity, 106
System volume, changes in, 13
Systems
  defined, 1
  in equilibrium, 1, 13

T

Tails, 91, 139
Tangential stress, 150
  consequences, 160
Temperature, 1
  and adsorption, 85
  beyond, 13–15
  and chemical thermodynamics, 13
  critical, 7
  dependence of velocity on, 182–183
  effects on proteins, 117–118
  enzyme sensitivity to, 185
  and gel formation, 109

and phase inversion, 153
and reaction constant, 44
and reaction rate, 182
and shape of self-assembled structures, 104
and thermal motion, 19
and volatilization, 35
Temperature and volume plots, 71
isobaric transformation, 4
isothermic transformation, 2
Temperature scale, defining, 3
Tertiary structure, in enzymes, 185
Tetravalent interconnection, 173
Thermal capacity, 24
Thermal denaturation, 120
Thermal equilibrium, 4
Thermal motility, 18
Thermal motion, 22, 35
random, 58
temperature and, 19
Thermally neutral reactions, 17
Thermochenistry, 16–17
Thermodynamic adhesion, 81
Thermodynamic systems, 1–2
pressure, volume, temperature
parameters, 1
three parameters, 1
Thermodynamics, 13
adsorption, 85
emulsions and foams, 130–131
irrelevance of time, 173
self-assembly, 100–104
of solutions, 35
Thixotropic materials, 168
defined, 164
flow curves, 164
viscosity of, 165
Time-dependent deformation, 168
Time-dependent non-Newtonian flow, 164–165
Time-dependent rheological phenomena, 168
Time-independent non-Newtonian flow, 162–163
Time perception, 157
irrelevance to thermodynamics, 173
Total energy, 14
Total free energy gain, protein folding, 117
Total pressure, 38
Trains, 138
Transformations, initial and final states, 13

Trapped energy
in emulsions and foams, 141
freeing on coalescence, 152
Triglycerides, 27, 59
crystalline structures, 28
kinetic energy, 30
liquid state, 29
melting point, 28
polymorphism, 28, 29
saturated, 39
solid state, 30
Triple points, 24, 25
Truncated cones, 100, 107, 108
Turbidity, 97
Turbulent flow, 142
Reynolds number, 143
Two-tailed molecules, 109

U

U-tube theometer, 171
Universal gas constant, 3

V

Vacuum-packaging, 24
van der Waals equation, 6–9
gas molecule schematic, 7
solutions, 8
triple solutions, 8
van der Waals forces, 9, 131–133, 134, 135, 136, 137
lyophobic interactions as, 137
van 't Hoff isochore, 45
Vapor pressure, 36, 37
azeotropic mixtures with two components, 40
change for two-substance mixture, 39
and partial pressure, 37
and Raoult's law, 41
of solutions, 39
Vaporization, 26
entropy of, 26
latent heat of, 23, 27
Variables of state, 24
Velocity
and chemical kinetics, 174
dependence on temperature, 182–183
and depth in Newtonian flow, 162
formation and breakdown, activated complex, 185
of second-order reactions, 180
Velocity-substrate concentration, 186

Virial coeffiients, 10
Virial equation, 9–10
Virial-like equation, 68
Viscoelasticity, 158
    creep method, 168
    determination methods, 168–170
    dynamic measurements, 169–170
    in gels, 168
    linear, 168
    oscillation methods, 169–170
    relaxation and, 169
Viscosity, 5, 35, 159, 161
    continuous phase, 149, 150
    intrinsic, 60
    linear relationship with volume
        fraction, 167
    reduction over time, 164
    sudden change in phase inversion, 153
Viscous behavior, 157
    Newtonian flow, 161–162
Visible light wavelengths, 147
Volume, 1
    gas for single layer coverage, 88
    relationship to temperature and
        pressure, 2
    undesirable increase in, 17
Volume change, 159
    products *versus* solvent, 15
Volume exclusion, 138. *See also* Excluded
        volume forces
Volume fraction, 106, 145–146, 165
    of dispersed phase, 166
    emulsion structure changes as function
        of, 166
    linear relationship with viscosity, 167

W

Water
    cousin molecule, 133–134
    dissociation of, 54
    protonation of, 54
    strong polarity, 91
    surface tension and, 77
Water–air interfaces, 123
Water–fat interfaces, 123
Water-in-oil emulsion, 128, 129, 147
    transition by phase inversion, 153
Water-in-water emulsions, 128, 129
Water molecules, ionic competition for, 118
Whipped cream, as emulsion and foam, 130
Wire frame experiment, 77, 78
Work
    mathematical expression, 13
    and surface tension, 78

Y

Yogurt, 170
    flocculated colloids in, 150
    increase in storage modulus, 170
    oscillation study of setting, 171
    transition of milk to, 170, 171
Young's modulus, 159

Z

Zero-order reactions, 176–177
    course of, 177
Zeroth law, 4
Zeta potential, 134

For Product Safety Concerns and Information please contact our EU
representative GPSR@taylorandfrancis.com
Taylor & Francis Verlag GmbH, Kaufingerstraße 24, 80331 München, Germany

www.ingramcontent.com/pod-product-compliance
Ingram Content Group UK Ltd.
Pitfield, Milton Keynes, MK11 3LW, UK
UKHW021120180425
457613UK00005B/161